Nature's OUTCASTS

Nature's OUTCASTS

A New Look at Living Things We Love to Hate

by Des Kennedy

A Storey Publishing Book

Storey Communications, Inc.
Schoolhouse Road
Pownal, Vermont 05261

Editor: Elaine Jones
Cover Designer: Meredith Maker
Cover Illustration: © Robin Brickman
Interior Designer: Carolyn Deby
Interior Illustrator: Sandy McKinnon
Typography: CompuType, Vancouver

Published in the United States in 1993 by Storey Communications, Inc.,
Schoolhouse Road, Pownal, Vermont 05261.
Originally published in 1992 in Canada
as *Living Things We Love to Hate*
by Whitecap Books, Vancouver/Toronto.

Copyright © 1993, 1992 by Des Kennedy

Library of Congress Cataloging-in-Publication Data

Kennedy, Des, 1945–
    [Living things we love to hate]
    Nature's outcasts : a new look at living things we love to hate / Des Kennedy.
        p.   cm.
    Originally published : Living things we love to hate.
Vancouver : Whitecap Books, 1992.
    Includes bibliographical references and index.
    ISBN 0-88266-868-4 (pb)
    1. Biology, Economic.   2. Pests.   I. Title.
QH705.K46   1993
574.6–dc20   93-182
               CIP

Printed in Canada

Contents

Foreword   vii

Preface   ix

Acknowledgements   xiii

Chapter One: Bats—The Devil's Bird   1

Chapter Two: Alders—Healing the Wounds   13

Chapter Three: Fleas—The Jokes Aren't Funny Any More   23

Chapter Four: Deer—The Terrible Beauties   33

Chapter Five: Slugs—Nature's Slimy Recyclers   45

Chapter Six: Dandelion—Tramp with a Golden Head   55

Chapter Seven: Pigeons—Rock Doves on a Roll   65

Chapter Eight: Flies—Awful Fecundity   73

Chapter Nine: Moss—Softening the Stones   83

Chapter Ten: Mice—From Steppes to Stars   93

Chapter Eleven: Raven—The Great Transformer   105

Chapter Twelve: Raccoons—Crafty Handiwork   115

Chapter Thirteen: Spiders—The Sinister Spinners   125

Chapter Fourteen: Starlings—Murmurations and a Mirror   135

Chapter Fifteen: Snakes—Saint Patrick's Malediction   145

Chapter Sixteen: Stinging Nettles—Devilishly Delicious   155

Chapter Seventeen: Rats—Who's Racing Whom?   163

Chapter Eighteen: Toads and Frogs—The Free Tenors   173

Chapter Nineteen: Wasps—The Social Terrorists   185

Chapter Twenty: Wood Decomposers—Dealing with the Debris   195

Metric Conversion Table   204

Bibliography   205

Index   210

Foreword

*S*how a schoolchild a slug, spider, bat or salamander and chances are, the child will recoil in horror, fear or loathing. Our attitude to "creepy crawlies" reveals, I believe, the reason why our species has put this planet on such a terrifying path of destruction. "Nature" has become an enemy to our species. "Don't touch that!" parents warn as children move to pick up something moving. It's "dangerous," "dirty" or "disgusting" they are taught by the words, body language and actions of adults.

For years, my two young daughters treasured three fire-bellied salamanders that my father gave them. Those lovely amphibians accompanied us as we moved back and forth between Toronto and Vancouver. Whenever the girls brought a new acquaintance home for the first time, they would invariably dive into the aquarium to show off their salamanders. Every child would recoil in fear or disgust, uttering "Yuk" with disdain. Fortunately, natural curiosity and my daughters' evident delight would usually overcome that initial response.

This is strange because any parent knows that a very young child seeing, for the first time, a flower or butterfly, or a spider or snake for that matter, does not react with fear. The predictable response is delight and curiosity and, usually, a move to touch the object. I believe that such a response is coded within our genetic makeup. We have a deeply embedded need to be with other species. After all, our distant hominid ancestors evolved on the African plains in the presence of many other species on whom we depended for our very survival. As social animals, we also have an innate desire to share our space with other beings. How else can we explain the child at play with a dog or cat, the response of people in old-age homes or hospitals to pets? It is expressed in the delight that gardeners take in rooting about in the earth and watching plants grow. We express a profound intimacy with other living creatures that is necessary to our well-being.

But today 80 per cent of us live in urban settings. In such a human-created environment, we foster an illusion that we control our surroundings through our great knowledge and technological power. Thus, we hear of our ability to "manage" our natural resources, of being able to "reforest" a clearcut old growth forest, of controlling or mitigating the impact of dams, mills and factories on the environment.

It is a dangerous conceit. Our ignorance of the biological and physical world of which we are a part is vast. Smithsonian scientist Terry Erwin estimates there may be as many as 30 million species on Earth and to date we have identified a mere 1.7 million. And that only means a dead specimen has been given a name. It does not mean we know anything about their distribution, habitat, life cycle, or interaction with other species. In Canada, of all insects collected, fewer than half have even been identified, and insects are over 90 per cent of all animal species. In the Carmanah Valley on the west coast of Vancouver Island, a new research platform built into the canopy of a Sitka spruce has already allowed the capture of species and genera heretofore unknown to science. Scientists estimate that, at most, 30 per cent of the arthropods in that valley have been collected and that the distribution of insects is often confined to microhabitats that are lost when forests are clearcut.

By fostering the impression that we know all we need to know to manage nature and by encouraging a distancing between us and other species, we diminish the sense of awe and respect that is needed to temper our attack on them. Des Kennedy's book is a wonderful turnaround from this attitude. He recaptures that sense of innocent delight in other species that we all had at birth and too often were taught to forget. Des introduces us to our next of kin, defined by Harvard's E.O. Wilson as "all the other species that we share this planet with and who are related to us through our evolutionary history and our DNA. To know our kin, is to learn to love and cherish them." Wilson calls this love of other species biophilia.

It's easy to become entranced with the big things—lions, grizzly bears, orchids or whales. But while the world Kennedy explores may be less obvious, the creatures he describes are every bit as remarkable, graceful, elegant and, yes, beautiful too. Des Kennedy expands the range of our biophilia in this timely and important book.

David Suzuki

Preface

*T*wenty years ago, my partner, Sandy, and I left the comforts of city living and well-paying professions to take up life on a wooded acreage on one of British Columbia's northern Gulf Islands. We made that move without any conscious realization of how profoundly estranged from the natural world we were. Drawn "back to the land" in order to live a quieter and gentler life than seemed possible in the hurly-burly of the city, we had no clear sense that we were inserting ourselves into a natural environment aswarm with organisms of which we knew virtually nothing.

Twenty years later, I still know virtually nothing, considering how much there is to understand about an ecosystem as richly complex as the temperate forest that surrounds our home. Strange cries echo through the woods at night. Untold numbers of unknown organisms are at work breaking down old snags in the forest. Why are the tree frogs singing? It has always been our instinct to tread lightly upon this little piece of earth, and as the years go by, each bringing a few more insights into the magical dimensions of the natural world, that instinct waxes stronger.

But so do the forces of destruction, and the sobering truth is that they move in an ignorance even more profound than my own. The big decisions—to tear natural systems apart, to disarrange natural harmonies—are largely taken in corporate boardrooms, sterile environments about as far from the natural world as one can get. CEOs don't go barefoot in the spring, but have no compunction about closing "the deal" to flatten and "develop" another piece of earth.

It's not only that we humans, while wreaking such havoc, are ignorant of our fellow creatures—in many cases we embellish our ignorance with fantastic elaborations of superstition and fear. Take the bat, for example—a perfectly harmless, indeed wonderfully beneficial creature. Yet popular imagination recoils from it as both repulsive and dangerous, not because of awful personal experiences, nor as a result of solid information, but because of a fictitious concoction of rumours about vampire bats, musty superstitions about witches flying around at night, and lurid prose from gothic horror novelists.

In researching this book I was intrigued to realize how many of the despised species I'd chosen to examine were at one time reviled as the familiars of witches—creatures which, superstitious folk believed, aided black witches in their mischievous work. Bats, mice, rats, spiders, snakes, toads, wasps—all were whispered to be in league with malignant old crones. It's interesting to ponder whether this association with evil resulted from or contributed to the bad reputations these creatures have. Either way, these familiars hint at a strong connection between alienation from the natural world and the systematic oppression of women for being in league with the dark forces of nature.

I recently read the following definition of pest: "any living organism objectionable to man." This mentality sees humankind not as part of a web of life, but in the same way the earth was once seen—as the centre of the universe, around which everything else revolves. Humankind's success as a species, according to this attitude, rests upon its ability to deal effectively with competing organisms. We aren't in competition just with one another, bullish capitalism tells us, but with all living things. Ultimately they are all "objectionable to man" unless they serve our purposes. In this context, the slaughter of the great buffalo herds and the passenger pigeon make the same good sense as the exploitation of indigenous peoples. Only the strong survive.

Ironically, the most striking characteristic of some of these despised species is their close resemblance to the human race. Here I mean the tough and aggressive, pioneering and "weedy" species—the rats and dandelions, the starlings, slugs and flies—which are every bit as good as we are at adapting to new circumstances, overpowering more specialized species, and multiplying even in desolation. These are the inhabitants of regions most degraded by human activity. Scientists now speculate that if human destruction of ecosystems continues, and with it the mass extinctions we're now accepting as commonplace, within the biological eye-blink of a few centuries these aggressive and adaptable generalists will have seized virtually all of the ecological terrain once inhabited by a spectacular diversity of species. Our actions or inactions today may thus condemn our inheritors to a biologically impoverished wasteland, in which the only wildlife to be seen are the creatures we despise as vermin.

I find that prospect intolerable. And I'm no longer certain that it isn't inevitable. If things are to change, even at this late hour, we will begin with

a wholesale change in attitude towards all living things, indeed towards life itself. Our task is one of reconnection with the natural world, including those parts of it that we have learned to fear and despise. We are required to cast off careless assumptions and time-twisted superstition. Beyond them lie the deeper truths found in the ancestral wisdom of tribal peoples. Above all will come the recognition that our gravest cause for fear is human avarice, for in this lies the destruction of the natural world as well as the impoverishment of humankind. I hope that this little book, besides providing you with some hours of pleasure, serves as one modest component in that massive but absolutely mandatory transformation.

Acknowledgements

Several of these chapters appeared originally, in altered form, as feature articles in *Nature Canada* magazine. I wish to thank *Nature Canada* editor Barbara Stevenson and former editor Judy Lord for their encouragement over the years.

I greatly appreciate Whitecap publisher Colleen MacMillan for enticing me into writing this book, as well as designer Carolyn Deby and marketing director Robert McCullough for their contribution to the publication of this book.

Thanks as well to editor Elaine Jones for her thoroughness and tactfulness in red-pencilling the manuscript.

Special thanks to Sandy McKinnon, a very talented young woman who managed to produce the book's illustrations between being valedictorian for her high school graduation and plunging into campus life at the University of Victoria.

I have a tremendous appreciation for David Suzuki, who generously took time from a hectic schedule to write a foreword for the book. All of us, I believe, owe David a debt of gratitude for his unflagging energy in defence of the natural world.

Lastly, I wish to acknowledge my life companion, Sandy, whom the reader will glimpse through these pages. In the writing of this book, as in all else, she has been a support, an inspiration and an abiding source of joy.

Chapter One

BATS

The Devil's Bird

*O*nce upon a time I lived as a pious
young monk in a rambling old
Catholic monastery in the eastern
United States. It was a forlorn, forbidding place, dark and secretive in its
austerity. As the great bell tolled one
evening, calling the community to
prayer, and we filed in silence down

the cloistral corridors, a sudden, disconcerting flash of black shattered our contemplations. A small bat was trapped indoors. It flitted and swooped around the chapel, down a corridor and back again. Venerable monks began ducking and cowering in very undignified fashion. Giggles could be heard from the young seminarians. The solemn piety of the cloister was beginning to unravel badly. Then, boldly, one of the more athletic seminarians undid his thick, black belt, waited for the bat, and with a deft snap of the belt, the way locker room boys snap towels at one anothers' buttocks, struck the tiny creature in mid-flight. It tumbled to the terrazzo floor and was quickly scooped away. At the rector's solemn rap on his lectern, we composed ourselves, rose as one and intoned the Latin verses of compline.

And that, in brief, is about how it's been between bats and humans. The relationship has been characterized by fear and horror for centuries.

In 1948 a pair of "vermin exterminators" entered an old bunkhouse in Coquitlam, B.C., just up the Fraser River from Vancouver. They carefully sealed all holes and exits in the building, and then released into it a killing cloud of cyanide gas. As the lethal vapours crept among rafters and into crannies, dying bats began falling to the floor. In all, five thousand bats trapped inside the building died horribly. "Newsman in Blood-Curdling Brush with 5,000 Flying Mice" screamed the next day's headline. Grisly photographs showed piles of dead bats, and florid copy described "this hot-bed of vampire life."

A few decades later we were treated to the celebrated case of a solitary bat trapped in the Philadelphia Spectrum arena, trying desperately to orient itself amid the bedlam of sixteen thousand roaring hockey fans. Eventually one of the self-styled Broad Street Bullies swatted the creature out of the air with his hockey stick. As it fell to the ice like a crumpled scrap of dark parchment, the fans roared bloodthirsty approval.

"Flying rodent," "devil in disguise" and "witches' bird" our peasant forebears called the bat. Humans have feared and hated bats for centuries, loathed them as harbingers of death, and cursed them as companions of witches. You might think we're more enlightened by now, but not long ago a prominent psychiatrist specializing in fears and phobias told the *New York Times* that bats symbolize "the mother as devouring ogre, the bad mother who is going to destroy something," hence the expression "old bat."

Anyone who has been startled by a bat knows there's something about them that sends an involuntary shudder down the spine. Once when tearing cedar shakes off an old shed roof, I encountered a small bat sleeping (or rabid! I thought). I recoiled in horror. I was afraid of it, more afraid than

I would have been of a hornets' nest or a fierce dog. I'm interested to know where that fear comes from.

* * * * *

We begin by not knowing much about the origins of bats, nor about how much they've changed in the 100 million years experts think they've been around. They're scantily represented in the fossil record—the first confirmed signs of them date back about 35 million years, but researchers speculate they're far older than that. It's thought they may have evolved from some tree-top glider like the modern flying squirrel, and that they've changed relatively little through that enormous span of time. "Bats were catching insects in the prehistoric evening," write John Hill and James Smith in *Bats: A Natural History*, "when the ancestors of man himself were but early primates no larger than the lemurs that occur today on Madagascar."

Of all earth's mammals, only bats and humans have managed to defy gravity. Where we rely on fantastic technology to get us airborne, the bat has developed a peculiar little body for the purpose. Its "wings" aren't really wings at all, but hands of grotesquely elongated fingers, between which are stretched two layers of kid-leather-soft skin. The wing membrane continues back to connect the hind legs and, in some species, the tail. This peculiar arrangement perfectly suits the bat's erratic, darting flight pattern. Bats aren't fast—scientists have clocked big brown bats at up to forty-one kilometres per hour; perhaps the swiftest of them all, one of the free-tailed bats of Brazil, is thought to be capable of sixty kilometres per hour—but as flyers they're very, very good.

A bat's body is generally furry, with the narrow hips and broad chest of a miniature body-builder. Its face seems alert and intelligent, though its eyes are small and chronically weak (but not blind, despite the popular saying). The quintessential listener, its ears are huge—so large in some species that, when flopped down, they cover the bat's whole face and hang over the edge like a tablecloth.

About 950 known species of bat flit through evening skies around the globe. Only the poles and a scattering of remote oceanic islands don't have them at all. The biggest are the flying foxes of Southeast Asia, formidable characters which can weigh over 1300 grams and spread their leathery wings more than 1.5 metres. The smallest is the rare hog-nosed bat, which is known only from two caves in Thailand. Nicknamed the "bumblebee bat," this little chap measures only thirty millimetres from head to tail and weighs next to nothing. It rivals the pygmy shrew as earth's smallest mammal.

Where I live, more than a dozen bat species are native to the woodlands of the Pacific Northwest, and I like the descriptive common names they have, like the western big-eared bat, the small-footed myotis, or the hoary bat with its striking coat washed with bright silver. The two species with which casual observers are most familiar are both called brown bats, though they're not of the same genus. Both the little brown bat (*Myotis lucifugus*) and the big brown bat (*Eptsicus fuscus*) are numerous and widely distributed throughout North America, and they've adapted more than any other local species to living among humans.

The little brown bat spreads its wings about twenty-three centimetres, and its nine-centimetre body is well covered with long, silky brown hair. The big brown bat isn't really much bigger—its body is about twelve centimetres long, with a rusty brown coat, and its black wings spread about thirty-three centimetres. Both are forest dwellers, spending the lazy days of summer snoozing in a hollow tree or cave, and hibernating in local caves all winter. But as more and more of their forest habitat is levelled—plantation tree farms, unlike forests, don't have many old hollow trees—both species have shown themselves opportunists, quick to adapt to new conditions, and many now hole up in eaves, attics and belfries, or under shingles or loose boards.

As summer dusk settles over our place, bats slip from their daytime hiding places and begin to flit about in search of food. We see them darting across the evening sky, and when the porch light is on, they'll swoop back and forth past it, snatching flying insects drawn to the light. Adapted to the tricks of city living, they'll feed happily on insects conveniently congregated in pools of light along streets and highways.

Efficiently quartering back and forth, zig-zag fashion, at about sixteen kilometres an hour, they capture flying insects using two ingenious adaptations. The first is simple enough—a clever net formed by cupping the membrane stretched between their hind legs. The hapless insect is simply swept into it and trapped there. The second, and vastly more complex, tool is the remarkable faculty bats have for orienting themselves and for tracking prey by using sound, called echolocation.

Naturalists were long amazed at the bat's capacity to ''see'' in the dark, but only relatively recently have researchers confirmed that bats possess their own internal radar. In flight they emit a rapid series of cries pitched up around five thousand vibrations per second, well beyond the range of human hearing. With the best set of ears of any land mammal, they pick up their own echoing vibrations, miraculously avoiding obstacles while tracking prey with tremendous agility.

Recent scientific reports indicate that both bats and certain of their prey are continually modifying and refining their tactics to achieve a temporary advantage in their deadly game of hunter and hunted. It's been shown that moths, a favourite food of many bats, can hear and identify echolocating bat cries and will respond with evasive plummets, loops and dives. The bats reciprocate with constant adjustments of their cry frequencies. It's believed that some moths may try jamming the bat's radar by emitting similar sounds. Dr. James Fullard from the University of Toronto is quoted in the press as saying, "The amazing thing is the sheer volume of information processing and profound neurological decisions handled by both bats and the moths, using a very limited number of nerve cells. They exhibit a degree of economy and sophistication that could be the envy of human aerial warfare strategists."

Having netted a victim, the brown bat either lowers its head and eats the trapped insect in fluttering midflight or, with larger prey, alights on a perch to de-wing and devour its victim. A big brown bat can easily catch 2.7 grams of insects in an hour. One analysis of stomach contents found a diet of 36 per cent beetles, 26 per cent flying ants and related species, 13 per cent flies and crane flies and a smattering of other insects of the night —stoneflies, mayflies, lacewings, caddis flies and scorpion flies.

* * * * *

As the insect-filled nights of summer begin to cool and grow longer with the approach of autumn, adult brown bats start entering breeding season, and in both these species breeding goes cheek-by-jowl with hibernation. In order not to freeze to death in winter, the bats must abandon their summer haunts for large caves which maintain a temperature around 4°C and relative humidity of about 80 per cent. By early November they've begun congregating at these "home" caves or abandoned mine shafts. Usually these are near the summer roosts, but they may be as far as several hundred kilometres away. Scientists think bats can find their way over such distances by identifying remembered topographic features. It's known that bats have an acute homing instinct—researchers have removed them more than two hundred kilometres from their home range and the bats have unerringly returned home.

Once at the winter caves, they cluster together, hanging by their toes from the ceiling, wrapped in cloaklike wings, in dense congregations of hundreds or even thousands. As bats in general go, these are small congregations, mere hamlets compared with the fantastic populations found in places like the U.S. southwest. Bracken Cave near San Antonio, for example, swarms with some 20 million guano bats in summer—by far the largest gathering of wild mammals found anywhere on earth. Sometimes brown bats will hibernate alone or in isolated groups, but never mixed with bats of another species. The big brown bat particularly has taken to hibernating in the attics and cellars of suitable buildings, often with catastrophic results.

For little and big brown bats the winter sleep is more of an extended dormancy than true hibernation. Their metabolism drops in the cool moist air, and they subsist on stored fat. In mild weather they may revive enough to shift about or even go for a sip of water. Though not really deep sleepers, it's said that bats generally can spend longer stretches of time in dormancy than any other mammal. Little brown bats have been observed to remain motionless for eighty-six days, and in one heartless lab experiment a big brown bat imprisoned in a refrigerator remained alive but motionless for almost a year before dying of starvation.

One of the abiding mysteries about bats is why it can take a female as long as eight months to produce one tiny baby, when similar-sized mammals, such as rodents, can produce dozens in the same time. Bats often mate in the late fall just before hibernation, while the male testes are functional. Sometimes during hibernation an amorous bat will become aroused and mating will occur. But the real morning-sex types may wait until they've slept for several months, then copulate in the spring before leaving their cave.

After copulation, females have a remarkable capacity to hold live sperm in the uterus and delay fertilization for perhaps the whole hibernation period. In certain species, the female may expel sperm obtained in autumn love-making and, in a spring affair, acquire new sperm which the patient male has himself stored all winter.

All of this fooling around normally results in the birth, some time between May and July, of a single tiny bat. Hanging upwards for a change, by her thumbs, the mother catches her newborn in the same membrane net she uses for hunting. The tiny bat soon clambers instinctively upwards through its mother's fur and clings to her chest near her teats while she swaddles it in her wings. For the first few days, the youngster will remain clamped sideways across its mother's breast, even while she's hunting. Thereafter, whenever she's away, the pup is left behind in a communal nursing colony.

Observers have long wondered how nursing mothers are able to locate their pups in huge nursing colonies such as those in the southwest where there may be several million youngsters, often packed tightly together, up to forty in one hundred square centimetres. It was once believed that lactating females would nurse any hungry youngster they came upon, but a recent study found that mothers are successful at nursing their own pups 83 per cent of the time. How do they manage it in a bewildering mass of babies that makes a Tokyo rush-hour subway look deserted? Experts say they use both voice clues and a distinctive scent with which each mother has marked her pup.

Summer nursing colonies of brown bats are composed mostly of babies and females, along with a few immature males. Adult males roost separately, sometimes alone. Within a month the young are weaned and have learned to fly. The adults reunite by late summer, sometimes in less-protected roosting sites.

Young bats will mature fully during the next winter's sleep and be ready to reproduce the following year. Barring calamity, a bat should live a good long life for a mammal its size. One little brown bat banded as an adult in Vermont lived another twenty-four years. An Indian flying fox caged in the London zoo holds the record for captured bat longevity—thirty-one years and five months.

* * * * *

So the bat, by and large, is a gentle and unassuming creature. Why then this tumult of human fear, superstition and hatred? "Bats have long been regarded as sinister, demoniac, and generally undesirable creatures," write Smith and Hill. "No other group of mammals seems so shrouded in mystery, mythical folklore and misinformation." What haunted house or castle would be credible without a ghostly colony of bats? And imagine what a hollow Hallowe'en we'd have with not a single sinister bat fluttering above the heads of ghouls and goblins!

Country folk living in the border counties of Wales used to think that malevolent witches transformed themselves into bats at night and flew into peoples' houses. Superstitious Sicilians believed the bat was a form of the devil; they'd protect themselves from evil by burning a captured bat to death while singing a song to it. In parts of the Sudan they call the bat *Bitabok*, the same name used for spirits and witches. In ancient Babylon it was believed that ghosts in the form of bats flew through the night air. In the folklore of many peoples a bat entering a house or flying three times around a house was a certain omen of death. Dour Irish folk called the bat *bas dorca*, meaning "blind death."

For Mayan Indians bats were creatures of great power, occupying one of the regions of the underworld, a kingdom of darkness through which the soul of a dying person must pass on its journey to the nether world. Ruling this dark kingdom was a savage Death Bat or Vampire Bat God.

Here at least mythology is grounded firmly in experience, for the vampire bat (*Desmodus rotundus*) of Central and South America is a frightening bit of business indeed. Not the huge and fearsomely fanged bloodsucker of horror movie fame, it's a small creature whose attack is both subtle and insidious. The vampire approaches its sleeping prey by night; domestic livestock and even humans are likely targets. Stealth is its weapon; often the bat will alight softly nearby and crawl onto the victim's body, looking like a large spider. Softly it moves and chooses a piece of exposed skin: the neck or leg of a steer, often the big toe of humans. From a tiny incision made with its scalpel-sharp teeth, the bat laps up blood. The attack is so softly and cunningly done that the victim often remains peacefully asleep. An anti-coagulant in the bat's saliva keeps the blood from clotting, and so the bat gorges itself. One observer described how a vampire licked blood steadily from a domestic goat for ten minutes, finally becoming so bloated with blood it couldn't fly away.

The mortal danger of the vampire bat lies not in its bite, but in the diseases it carries. The worst of these is the terrifying paralytic rabies virus, transmitted through saliva. Despite widespread inoculation programs, it's estimated that something like a million cattle die every year in Latin America from rabies transmitted by vampire bats. Human fatalities also still occur.

Responding with more firepower than intelligence, humans have dynamited vampire roosting caves, fumigated the caves with poison gas and even attacked the bats with flame-throwers! Predictably, none of these tactics has proven very effective, and some have inadvertently killed many benign bats. Realistic solutions, say the experts, lie in biological controls, including sterilization, habitat management and other measures. Ironically, the number of vampire bats is believed to have risen in response to the millions of domestic animals now ranging on former forest land—yet another unexpected and unwelcome result of human destruction of tropical rain forests.

* * * * *

The terrifying symptoms of rabies infection—disorientation, random viciousness and foaming at the mouth, followed by docility, convulsion and death—quickly became part of general bat mythology, particularly with nineteenth-century gothic horror writers. Some of today's horror merchants in the news media love rabid bat stories too, and the last few decades have seen any number of rabies scares in the Pacific Northwest and elsewhere.

But how realistic is the fear of rabid bats outside the tropics and subtropics? Rabies scares will sometimes flare up when one or more bats is found on the ground in daylight. One should never touch such an animal, but the bat's disoriented state is more likely the result of some bungled poisoning effort than of rabies. While the rabies virus has been isolated in twenty-four of the forty bat species living in the U.S., it's believed to be no more common in bats than in any other mammal population; dogs and cats are more likely carriers. A few years ago a study concluded that over a thirty-year span only nine Americans and one Canadian had died from rabies contracted from a bat bite. "In fact," say Hill and Smith, "more people have died annually as a result of attacks by dogs, bee stings or lawn mower accidents than this thirty-year total of deaths through bat rabies."

Besides the rabies virus, bats carry other potentially harmful viruses, fungi and bacteria; but again, verified cases of human contamination are extremely rare. There have been isolated cases of people contracting the respiratory disease histoplasmosis after entering dusty bat caves where there's lots of guano. Labourers digging guano for commercial fertilizer can certainly be at risk. One researcher has concluded that the legendary Curse of the Pharaohs, mysteriously dealing death to those who would disturb the ancient royal dead, is actually a bat dung disease in the pyramids.

* * * * *

One summer afternoon I was hammering away on an outer wall of our house, nailing up some long-delayed battens. Perched on my ladder, I heard a frantic squeaking, and right before my face there emerged one, then two, then a third small bat. Each in turn squeezed itself through a small gap between the cedar boards and flew off into the woods. They're under the cedar shakes on our roof as well, and in the attic—we hear them scrabbling about and squeaking just before dawn in the summer.

Besides the nuisance of their noise, their accumulated droppings can have a certain aroma. But they don't do any structural damage and we find, on balance, we'd rather have them with us than not. In a more extreme case, a few years ago a condo in Oceanside, California, was invaded by hundreds of bats, hanging from every fixture. The terrified occupants fled into the night. "It looked like an Alfred Hitchcock movie," the shaken proprietor told reporters.

Banking on people's fears, exterminators have turned a tidy profit eliminating bats. For years DDT was used for the job, until its use was restricted. Still, as recently as 1988 the government of Ontario was issuing special permits to exterminators allowing use of this banned substance against bats. In a paper entitled "Bats and Public Health" Dr. Merlin Tuttle and Stephen Kern wrote: "Poisons used in bat control may seriously threaten humans as well, and cause sickened bats to fall to the ground, where they die slowly and may be picked up by inquisitive children or pets. Recently a single application of a toxicant resulted in a 700 per cent increase in human contact with bats."

Short of poisoning, one can attempt to drive bats off, and commonly used repellents include naphthalene flakes or paradichlorobenzene (PDB), both of which spread toxic vapours through the building. Floodlights, sticky paste on roosts, electric fans, high frequency sounds and smoky sulphur candles have all been tried as well, usually with limited success.

The best bet, everyone agrees, is to deny access to unwanted bats. Easier said than done, of course, for they can slip through the tiniest cracks, as I discovered with my batten hammering. Entrances should be blocked in spring or fall, so as not to trap young or hibernating bats inside, and at dusk when bats are out hunting. You're supposed to re-open the entranceways briefly at dusk a few days later to allow any stragglers to escape, then seal them permanently.

* * * * *

I like the approach in Great Britain, where recent legislation has made it a crime for anyone without a license to disturb roosting bats, to kill or capture a bat, or even to have a dead bat in your possession. This stiff legislation is in response to plummeting bat populations throughout Europe. In the United States, the Endangered Species Act lists four species of bats—10 per cent of the total species list—as endangered. Other populations are at risk as well. The Animal Welfare Institute's *Endangered Species Handbook* warns that many of the earth's 950 species are quickly becoming endangered. Hill and Smith write: "Perhaps the most insidious factor threatening world bat populations is the continued use of persistent and bioaccumulating organic pesticides such as DDT. . . . recent studies of massive die-offs of nursing baby Mexican free-tailed bats at Carlsbad Caverns suggest that these deaths may be linked to the detrimental effects of pesticide residues on the developing nervous system of these young bats."

It's said that the Los Angeles basin was once home to untold thousands of bats, many of which have now disappeared. And that story is repeated over and over. In our forestry and our relentless urban sprawl we cut down the old hollow trees in which bats roost. We level old barns and warehouses where once they hung from rafters. In quarrying rock we destroy their winter caves. We deliberately poison unwanted bats and unintentionally poison others as well. Then, to control crop-destroying insects upon which the missing bats once fed, we invent ultrasonic transducers to produce the sound of an echolocating bat and thus upset the insects' feeding and egg-laying habits. And we have the audacity to think ourselves terribly clever!

But there are ancient and wise traditions concerning bats, and perhaps in these we will come to recognize our folly. To the Chinese the bat symbolizes long life and happiness. The common "Wu-fu" figure of five bats encircling a tree indicates the five blessings of wealth, health, love of virtue, a long life and a peaceful death. The flamboyant Gypsies of central Europe saw the bat as a magical creature who might bring them good luck, and often a bat's bone was worn around the neck as a lucky talisman. For certain Australian aboriginals the bat was a sacred totemic creature. Sir J. G. Frazer, in his classic *The Golden Bough*, described the functioning of the sex totem: "For each man believes that not only his own life but the lives of his father, brothers, sons and so on are bound up with the lives of particular bats, and that therefore in protecting the bat species he is protecting the lives of all his male relations as well as his own." (Females had their own totemic creatures.) A trickster, a comedian, a clever fellow, half-animal and half-bird, the bat flutters through the collective memories of tribal peoples.

Sometimes, when we'd turned out the bedroom light and opened the unscreened windows on warm summer evenings, the voluptuous scent of honeysuckle would waft in, and with it there would occasionally come a soft fluttering through the darkness. We'd dive beneath the blankets, fearing a vicious bat attack, or the proverbial bat entangled in our hair. (That doesn't happen, by the way; Jean Harrowven tells how, in 1959, an English earl proved it by purposely placing three bats in three different ladies' hair, and each instantly extricated itself!) In our case, the bat would flutter about the bedroom for a bit and then depart. Now our windows are screened and we no longer have that ghostly evening visitor. But hearing them in the attic and seeing them swooping after crane flies by the porch light still sends the imagination spinning back across 100 million years and the heart to hoping that, despite the evidence, things may yet turn out all right for both of us flying mammals.

Chapter Two

ALDERS

Healing the Wounds

"A s far as I'm concerned," the old logger said to me, "alders are nothing but damn weeds, and the best thing you can do with 'em,"—and I think he spat a squirt of stereotypical tobacco juice on the ground here—"is knock 'em all down!"

I nodded in agreement. New to the

woods, a greenhorn, tenderfoot, city kid—who was I to argue? We'd only just taken up the homesteading life on one of B.C.'s Gulf Islands, and I knew nothing of alders. With hard hat and chain saw and hands calloused from a lifetime's hacking in the bush, the old guy spoke with an authority that would suffer no contradiction. Alders should be cut down. End of discussion.

So I got right to it. The first big tree of any sort that I ever felled was a dead red alder. Its trunk snaked up for maybe fifteen metres in the middle of a debris-choked clearing on our land. Freshly arrived from Vancouver, bursting with back-to-the-land bushwackery, we intended to construct a small cabin in the midst of the previous owner's logging slash, and for safety's sake the solitary alder must come down. I fired up the chain saw—a screaming beast of terrifying potential we'd only just acquired. With a knot of city friends standing at a safe distance and prepared to shout "Timber!" at the tree's first sign of movement, I set the chain saw chewing into its trunk.

This was unnerving work, trying to complete an undercut the way the books advised, all the time glancing up at the menacing tower of wood above and imagining "widow-makers" and "barberchairs" and the other deadly tricks we'd read a falling tree can play. I roared into a shaky back cut and the tree began to list. Dropping the saw, I ran for my life! Everyone was screaming. The tree held, suspended for a moment by invisible skyhooks, then creaked like a campy horror-movie door, crashed to earth and exploded into fractured chunks. I resumed breathing and wiped my sweaty palms. The danger past, I puffed my chest ever so slightly and swaggered back to see what a wonder I'd wrought.

This was an admittedly modest beginning in the eradication of alder to which I'd been urged by the old-timer. Nor was he alone in the urging: at about the same time, unknown to me, a prominent botanist at the University of British Columbia was warning foresters that the uncontrolled spread of red alders represented "an epidemic of trees" which demanded prompt and effective countermeasures. Throughout the Pacific Northwest, researchers and foresters were conducting field trials, writing scientific papers and attending symposia on what to do about the alder epidemic.

These white-stemmed woodland invaders are red alders (*Alnus rubra*), also known as Oregon alder. They're the largest member of the genus, but not the only one that is despised as a "weed tree." The speckled alder (*A. rugosa*), also called grey or hoary alder, ranges throughout the eastern part of the continent and through the northern boreal forest. In rural areas of the east, where marginal farmlands have been abandoned, speckled alder creeps into damp and low-lying pastureland, multiplying to form dense thickets of squat, crooked shrubs. To those who remember fields once abundant with crops and animals at pasture, the alders are a blight, an infestation symbolizing failure, abandonment and the collapse of rural economies.

There are about thirty species of alder widely distributed throughout the temperate and cooler regions of North and South America, Europe and Asia. Members of the birch family, many of them grow as shrubs, but others, including red and speckled alder, can grow into sizeable trees. Three different species that thrive in the mountainous regions of western North America are commonly called mountain alder. They are the white alder; the Sitka, or green, alder; and the true mountain alder, which some people call thinleaf alder. All three are moisture-loving mountain dwellers, flourishing in canyon bottoms and alongside rushing streams, around high mountain meadows and lakes. Together they range from Alaska to the American southwest and as far east as Saskatchewan.

Besides these natives, and several others that occur in pockets across the continent, a number of exotic species have been introduced for landscaping purposes. These include the Italian alder, a native of southern Italy with a rounded crown and dense, glossy foliage. The European, or black, alder has been introduced for wetland planting. A third import, the Manchurian alder, is a handsome, pyramidal tree native to Japan and Manchuria. But even these introduced species carry a burden of contempt. "All are rather weak-wooded, susceptible to several pests, and most will survive only in wet soils where nothing else will grow," warns *Wyman's Gardening Encyclopedia*. "If plantings must be made in such situations, the alders are recommended, but if better soils are available, better ornamentals should be selected."

* * * * *

Until about 1930, foresters regarded alder as a weed and didn't even bother recording it in their timber cruises. But long before them, when forests were seen as something more than fibre farms, a different attitude prevailed. In 1630 the poet William Browne wrote of

> The alder, whose fat shadow nourisheth
> Each plant set neere to him flourisheth.

And in 1882, an official of the California State Mining Bureau, named Kellogg, produced a book called *Forest Trees of California* in which he wrote affectionately of *Alnus rubra*, which is found as far south as Santa Barbara.

They also maintain their broad leaves so perfectly horizontal, and the spreading branches so nearly so, as to afford one among the finest, most open, and aeriest of canopies—what was designated of old as the dense "fat shadows," beneath which the green grass and the tender herb continued to flourish ... Like the willows, their multiplied shallow roots preserve margins from the wear and tear of aggressive streams, and during the latter portions of the year, shelter, cool, and sweeten them, and altogether with the falling leaves, infuse and tone sluggish and stagnant waters ... The leaves are of some repute as fodder, the bark for tanning, and with twigs, tags and young wood, as a tonic, in teas, beers, etc.; for diseases of the skin, as detersive and expectorant, and a gargle in ailments of the throat; for ointments, etc.

The medical uses Kellogg mentions were picked up from native medicine people who used alders for a variety of cures. In 1672 one John Josselyn reported that "an Indian, bruising and cutting of his knee with a fall, used no other remedy than alder-bark, chewed fasting, and laid to it; which did soon heal it." Many of the tribes boiled alder bark to make a tea which was drunk to ease cramps and retching, to cleanse "impure blood" and to cure constipation. A bark tea was also employed as a vaginal and rectal douche, to combat fevers, dropsy and gangrene and as a decoction against intestinal worms.

Natives squeezed a juice from the slimy cambium layer found beneath the outer bark and rubbed it over the skin to cure itching. West Coast peoples ate the cambium layer, either fresh, mixed with oil, or as a dried cake stored for winter. After Europeans arrived, some bands mixed sugar into the pulp. Red alder bark juice makes a beautiful red dye, which natives used to colour cedar bark strips for weaving. Some coastal peoples spread bark in steaming pits with camas bulbs to colour the edible bulbs red.

A tree of cool, moist places, alder has an uncanny knack for cooling and soothing. As Josselyn reported, it's "excellent to take the fire out of a burn or scald." A few fresh alder leaves placed inside shoes relieve the discomfort of swollen, tired and hot feet—as though one's aching feet were being immersed in a cool mountain brook along which alders grow.

* * * * *

None of this lore was foremost in my mind when, back on the island, with cabin complete, we set about clearing a bit of land for house and gardens.

The area we needed to clear was covered with a dense grove of mature red alders. I have since come to appreciate what beautiful trees alders are in all their growth stages, especially in their prime. Their clear, straight boles, smooth and slender as adolescent limbs, stretch skyward with a sinuous, almost muscular grace. High above, their branches spread into an airy crown of slender, shiny twigs and thinly spread green leaves through which sunlight flickers. Their bark is light grey to almost birch-white, thin and smooth, though on old trees roughened with small swellings, cracked and black-patched and mottled with mosses. In late winter the small brown female "cones" and pendulant red male catkins hang in rusty clusters against the slate grey sky.

In the grove we proposed to clear, the alders stood nearly thirty metres high, with trunks fifty centimetres thick at their base. That's about as big as alders get on a prime growing site. Underneath them was a lush, chest-high ground cover of sword ferns. There was no bulldozer on the island when we started—or perhaps we couldn't afford one—so we dug out the ferns by hand, dragged their huge rootballs out of the clearing and replanted them in the surrounding woods. That left us with several dozen large alders to deal with. Cutting them down and digging out their roots by hand was out of the question; I'd already dug out several big stumps using a mattock, pry-bar and come-along, and while this is a wonderful exercise in appreciating the structures and strengths of roots, it's no way to clear a house site.

Instead we developed our own "big bang" approach—blow the trees out of the ground, roots and all! Off we went to a local explosives supplier and loaded the pick-up with enough dynamite to start an effective insurrection. We were given a five-minute training session in how to insert a fuse into a detonating cap and crimp the cap on, then insert the cap into a dynamite stick and combine several sticks with a bag of explosive powder. Simple. Off we went. After blowing out a small cavity under the roots of a tree with a single stick of explosive, we'd pack a full charge into the hole, light the fuse and run like hell to hide behind a big tree in the woods. The wait was agony—was it going to blow or not? Sometimes it didn't; we'd have to wait awhile, tiptoe up to the tree and whip the burned fuse out without blowing ourselves to pieces. But when it worked properly the ground would shake with a thump and the great tree would leap into the air and come crashing down, levering its root mass out as it fell. Whoopee! We cleared that bit of land all right, and now, looking back, I shake my head and file the whole adventure in my already bulging "too dumb to know any better" portfolio.

* * * * *

Our amateur explosives, however, were nothing compared to the onslaught against alders being waged by governments and forestry companies. As the full magnitude of the timber supply crisis became apparent in the 1980s, forest managers scrambled to increase timber production by flattening stands of red alder and planting conifer seedlings. Standard "site preparation" in the Pacific Northwest prescribed cutting down the alders and burning them, spraying the area with herbicides to minimize resprouting, then planting conifers and maintaining them in a "free-to-grow" situation by subsequent herbicide sprayings to kill off new alder or other "weed" growth. Millions of tax dollars and millions of litres of herbicides have been poured into this madness. But there was no victory here, and increasing public opposition to herbicide spraying, slash burning and monocultural tree farms may have doomed this battle plan completely.

What frustrates foresters and renders their efforts against alders so often futile is the tree's knack for pioneering. The alder is what population biologists call an "R-Strategist"—an opportunistic species, adept at discovering available habitat and reproducing rapidly to exploit its resources before another species can do so. A prolific seeder, its light seed can fly on a breeze for many kilometres. The seed needs mineral soil in which to germinate and does best in large openings where the earth has been recently disturbed. Clearcut logging sites—where the ground has been chewed up by heavy equipment, compacted in landings and roadways, and scraped over at sidecuts—are open invitations to wind-blown alder seeds, which drift in by the thousands.

"Red alder is a community tree, in that it requires a large number of its own kind to successfully establish and grow well," says Peter Sanders, resident silviculturalist at the University of British Columbia research forest in the Fraser Valley. "When it seeds in, in a pioneer situation, it'll come in at fifty thousand seedlings per hectare, and there'll be intense competition over the next five to seven years." On a favourable site, a red alder seedling can grow as much as forty-five centimetres in its first year, by its fifth year be over five metres high, and have reached twelve metres within ten years! "It peaks out really early," Peter Sanders tells me. "It culminates in biomass production at about age seven." By then, 95 per cent of the original seedlings are dead, squeezed out in the fierce competition for light, moisture and nutrients by about four hundred survivors per hectare.

Living fast and dying young, some red alders begin showing dieback and other aging symptoms as early as age forty. On certain sites, the canopy starts breaking up by age fifty, and even under best-case conditions few stands will survive for a hundred years. The big trees I blew out of our house site were in their prime of life, and when we counted the growth rings in one of the biggest, we were amazed to find it was only about forty years old.

Preferring to grow tightly packed together, and collectively weaving a dense canopy of branches, a stand of red alder can totally dominate a site to the exclusion of almost everything else. Certain understorey species like salmonberry and sword ferns have adapted to the mottled "flicker light" produced by sunlight dappling through the canopy, but few conifers will survive there, and fewer still can compete with alders in their prodigious early growth. Douglas-fir reaches its maximum rate of growth ten years later than alder, and if the two of them begin life at the same time, alder quickly outgrows and dominates the Douglas-fir. After an alder canopy has closed, understorey Douglas-fir suffer a marked decrease in growth and often die within seven years. Even more shade-tolerant species like hemlock may remain badly suppressed beneath aggressive young alders.

Companies engaged in intensive timber cropping naturally take a dim view of weedy alders suppressing more valuable conifers. But times are changing, and a new generation of foresters seems better prepared to include in their management plans the vital ecological role alders play in forest succession.

Among the tree's best-known and most valuable contributions is its capacity to fix nitrogen in nitrogen-deficient soils. Alder roots contain clusters of nitrogen-fixing nodules like those found on legumes, and additional nitrogen gets added through leaf litter and root sloughing. Newly developing soils exposed by recent glacier retreat and planted with alders show these pioneering trees are applying the equivalent of ten bags of high-nitrogen fertilizer to each hectare per year. Other chemical changes to soil in which alders are growing include a lowering of the base content and rise in soil acidity, as well as an addition of substantial carbon and calcium to the soil. "Since ancient times," writes forest researcher Robert Tarrant, "alder has been recognized as a soil-improving tree in Aomuri district [Japan] where cutover alder sites are usually utilized as farmland."

Another important role many alders play in the wild, and particularly in mountainous areas, is to check the rush of water during spring melt. In Japan and elsewhere, the trees are planted to stabilize soil on steep mountain slopes. Similarly, alders have been planted to stabilize and rehabilitate mine spoils, flood deposits and landslide areas in both Europe and Asia. Hundreds of thousands of hectares of strip mine spoils in the eastern U.S. have been rehabilitated with European and speckled alder.

Other benefits have been discovered; for example, alders have been shown to resist the root rot fungus *Poria weirii* which is one of the major root pathogens on conifers in western North America. All alders provide important habitat for wildlife. The rapidly decaying wood of red alder hosts multitudes of decomposer organisms and the creatures that feed on them. Speckled alder, particularly, provides important habitat for gallinaceous birds such as grouse, ptarmigan, quail and pheasant. Denizens of moist woodlands, swamps and thickets, including the American woodcock and the common snipe, rely on water-loving alders too, and its thickets make important nesting and roosting places for several dozen species of songbird. Rabbits, hares and other mammals browse on alder leaves and twigs.

Peter Sanders echoes Kellogg's praise of a century ago, when he says, "Along creeksides alders make a very nice overstorey for shade, and in combination with other species produce a good mix of insects for fish feeding." In many areas, beavers rely upon pondside alders for their food and building materials. While they prefer aspen or willow as food, beavers can live on alder leaves and bark alone. At a beaver pond in the woods behind our place I watched, over the course of several weeks, how beavers felled an enormous red alder, chewing through a trunk that was at least fifty centimetres in diameter. After it had crashed into the pond, they methodically stripped it of leaves, twigs and branches.

More than a decade ago Robert Tarrant advised foresters to "consider alder, not as an aggressive weed, but as a tree with considerable potential in a more mature forest economy." For example, maintaining a component of alders—perhaps forty to sixty trees per hectare—has been shown to improve conifer growth on nitrogen-deficient sites. Similar results have been obtained mixing red alder with cottonwood. Along the road where I live, a new plantation of Douglas-fir seedlings benefitted tremendously by getting established under the protective light shade of alder saplings which were manually removed after about three years. In eastern Ontario, forest researchers have been experimenting with interplanting four different alder species in hybrid poplar plantations to supplement the high nitrogen requirements of poplars. European foresters have already demonstrated that underplanting of alders in softwood plantations results in bigger and healthier crop trees.

* * * * *

Alder wood itself draws mixed reviews—for some uses it's very, very good, and for other things it's hopeless. At home we use it for a large portion of our winter's firewood. On cold, clear winter days I select a number of big trees in the woods—an examination of the crowns shows which are already starting to die back and become dangerous. Felled now with a confidence entirely lacking in my first tentative stab at lumberjacking twenty years ago, I still take extreme care when falling the big trees. Their rapid and eccentric growth patterns, where they literally twist and lean in the mad contest for sunlight, can put tremendous growth stresses into the wood. I've had them "barberchair," where a whole section of the standing trunk snaps loose with a force that can flatten your face. Falling branches are a constant menace, and, once down, big alders can be fearsome "sweepers": when the whole trunk, wedged under tension between standing trees, suddenly releases and "sweeps" sideways with bone-crunching power.

The wood itself—fine and white, close-textured and straight-grained, clean, uniform and smooth—splits beautifully for firewood. I stack the split pieces in piles and leave them to air-dry over summer. By autumn the pieces are far lighter and have coloured a dull reddish yellow. In the wood heater, it burns steadily without sparking or creosote problems, producing only moderate smoke and ash. The wood makes for mid-range firewood—not nearly as rich in BTUs as oak or birch, but better than lodgepole pine or aspen.

Light, soft, porous and smooth-textured, alder wood is quite fine for furniture, cabinets and wooden ware. It glues together with a strong bond, and holds paint and enamel well. During the Great Depression, people carved wooden shoes from alder. Coast natives valued the wood for carved dishes and spoons, as well as rattles, headdresses and masks. A skilled carver could shave the shell of an alder rattle as thin as two millimetres and polish it smooth with abrasive sharkskin.

Fine material for the splitting maul or the craftperson's blade, alder is less cooperative in the highball world of industrial logging. For one thing, you have to mow down a lot of alder to glean a little bit of good-quality wood. One study found that in a typical stand of mature alder, 35 per cent would be left behind as tops, stumps and cull material, and another 16 per cent be later discarded as bark and sawdust; about 35 per cent would make satisfactory pulp logs and chips, while only 12 per cent would be good enough for dried lumber for the furniture industry. Unless dried quickly, the wood suffers sap stain, and sawmill operators talk about how stresses in a log will sometimes cause the whole log to jump out of a saw carriage. Similar problems limit its use in pulp and paper mills—the knobbly logs are difficult to debark, and chips can't be stored for any length of time before they begin decomposing. Lacking the strength of softwood fibres, alder is used as an extender in certain kinds of pulp. Given these limitations, it's not surprising that the alder market is highly volatile.

I don't think anyone foresees plantations of alders replacing today's plantations of conifers or other commercial species. But many observers believe these remarkable trees may be well adapted to future requirements for fast-growing fibre for pulping and fuel wood. As the ground swell of public revulsion against excessive clearcut logging continues to mount, the role of alder as a necessary part of forest succession is apt to be more widely recognized. Perhaps the brutally scarred forest lands of the Pacific Northwest, the spoils of strip mines in the east, and other places laid waste by human activities may once again be healed and their soils restored by this lovely and misunderstood "weed."

Chapter Three

FLEAS

The Jokes Aren't Funny Any More

*A*fter centuries of stardom, the flea's career as a world-class entertainer has crashed. But what a career it was! Few stars have glittered more brilliantly in the firmament of fame; in its glory days flea was as sexy as Monroe, as funny as Chaplin and as acrobatic as the Flying Wallendas. The acknowledged star of fables and poetry, flea had songs and indeed whole operas dedicated to its brilliance. Entire circuses flourished with fleas alone, requiring neither elephant nor clown to supplement their multiple talents. Jokes about fleas were on the tip of every comic's tongue (Say, didja hear the one about two fleas who were goin' on a trip, and one flea says to the other: "Shall we walk or take a dog?").

And somehow flea was found in the naughtiest of stories too. "I am like Ovid's flea," Marlowe has one of his characters leer in *Faust*, "I can creep into every corner of a wench." How often prurient Victorian pornographers would picture a blushing maiden searching her person for a bothersome flea, and have her seducer lend a hand in its pursuit, removing one bit of concealing clothing after another, until . . . oh, shocking!

In its amorous adventures flea knew neither decency nor decorum. A shameless voyeur, it would peep across the ridge of an exposed buttock to watch the grotesque grapplings of human lovemaking, or peer close-up from a strategic thicket of pubic hair. "They follow their patrons from barrack to bordello, from ballroom to four-poster, from spacious hall to grimy hovel, and they have a particular zest for monasteries and nunneries." So writes Brendan Lehane in his delightfully definitive book, *The Compleat Flea*, to which I am much indebted concerning fleas.

Lehane in turn acknowledges his indebtedness to one Signor Bertolotto, the Andrew Lloyd Webber of flea-biz. With entrepreneurial shrewdness and tactical sagacity, Bertolotto lifted the performing flea from the rustic obscurity of the country fair and made it an international star. "Extraordinary Exhibition of the Industrious Fleas," proclaimed one of his posters in the 1830s. Royalty and the nouveau riches flocked to his theatre in London's Regent Street, paid their shilling and were handsomely entertained by a repertory company of performing fleas in period costume. The program notes, writes Lehane, "tell of a ball at which flea ladies partner their frock-coated gentlemen, and a twelve-piece flea orchestra plays audible flea music, while in an alcove four whiskery old flea bachelors make up a four at whist."

Signor Bertolotto's performers were not, he assured his patrons, just anonymous faces in a crowd of fleas. "By constant practice," he wrote, "I know my own fleas as a shepherd knows his cattle." Following the grand ballroom scene, the maestro brought on four fleas pulling a coach complete with liveried flea coachmen. Then came a flea harem scene and, lastly, the heroes of Waterloo. The crowds loved it. Bertolotto and his fleas caused a sensation, and the glamour they created shone for a century, its last glimmerings finally fading completely in the 1950s with the closing of the last tattered flea circuses.

* * * * *

A shocking scientific exposé dealt the eventual death blow to flea's show-biz career; but before the Age of Science dawned, mystics and moralists, poets, painters and philosophers mused mightily upon the flea. Three characteristics were obvious: its tiny size, its vexatious bites and its astounding capacity to leap. Flea's proper name, *Pulex*, is from the Latin word for dust, and ancient Romans believed the little villains to be animated bits of dust or dirt. Centuries later a French poet referred to them as, "a speck of tobacco with a spring in it." Tiny but persistent, flea became the inspiration for many aphorisms: thus, "a mere flea bite" is a trifling inconvenience or discomfort; "to flay a flea for hide and tallow" is to behave like an obsessive miser; and "to turn a flea into an elephant" is to make a big deal out of very little.

The flea's fantastic jumping capacity has long excited human interest. A U.S. Public Health official carried out a battery of tests in 1910 and discovered that among all fleas the premier jumper is the common flea (*Pulex irritans*), capable of jumping about twenty centimetres high and thirty-three centimetres lengthwise. Somebody else calculated that if a man, proportional to his weight, could jump as high as a flea, he could easily clear Saint Paul's Cathedral (110 metres high). Propelled by the elastic energy in its jumping legs, the flea often somersaults head-over-heels in the air, its clawed legs splayed out for landing.

Not only does the flea achieve this prodigious jump, 130 times its own body length, but it does so repeatedly. One study recorded how ten fleas, stimulated by each other's presence, executed ten thousand jumps per hour—one jump every 3.5 seconds. The oriental rat flea (*Xenopsylla cheopsis*) has been observed to jump as much as six hundred times an hour for seventy-two hours without a break! No wonder Signor Bertolotto's indefatigable performers, constrained by tiny daubs of cement and harnesses of twine, could do fifty ten-minute shows every day for two weeks!

All this hopping is for one purpose—to get the flea on a suitable host, where it can feed and breed at its leisure. How effective the "hop 'til you stop" approach can be was demonstrated by an experiment which Lehane cites. Researchers scattered 270 fleas throughout a field of some 17,000 square metres. Three flealess rabbits were introduced. Within days, almost one-half the released fleas had located and settled onto the rabbits.

"I think this be the most villainous house in all London for road fleas," complains a Shakespearian carrier in *Henry IV.* It's not surprising that in Denmark, India and elsewhere, folklore has it that fleas were sent by God to punish humans for their laziness and to goad them into action. If true, there are some categories of people who require more punishing and goading than others, because fleas have definite preferences concerning the age, gender and fleshiness of their hosts. Likewise, there's a spectrum of reactions to flea bites. The irritation is caused by an anticoagulant saliva the flea injects while feeding. Some people seem completely immune to it, some develop immunity after repeated bites, while others become increasingly susceptible. Generally speaking, young people seem more susceptible than old, and women more than men.

* * * * *

For centuries seen as an irritating but somehow charming little clown, the flea is no longer funny. Today we know it as the most efficient killer in human history. The seventeenth-century mystic and poet John Donne wrote, "The Flea, the Flea, though he kill none, he does all the harme he can." The line was uncannily prophetic. William Blake once had a vision in which a flea appeared to him in the form of a muscular and sinister-looking man. Depicted in Blake's painting, *The Ghost of a Flea*, the eerie apparition told the poet that all fleas were "inhabited by the souls of such men as were by nature bloodthirsty to excess, and were therefore providentially confined to the size and form of insects; otherwise, were he himself for instance the size and form of a horse, he would depopulate a great portion of the country."

Ironically, depopulating the country is just what the flea had been doing for the previous three centuries, though neither Blake nor anyone else knew it. Beginning in the fourteenth century, the Black Death spread remorselessly across Europe, killing some 25 million people, a full quarter of the population. In London alone over 200,000 people died, their corpses carried off in the dreaded pest-carts. With pest-houses full of the sick and dying, diarist John Evelyn described, "multitudes of poore pestiferous creatures begging almes."

Blamed for causing the plague, Jews were tortured and killed. Suspected witches and other innocents were burned, dogs and cats slaughtered.

Meanwhile, the true villain went about its deadly work unnoticed. The Black Death was the second of three huge pandemics documented over the past two thousand years. In the sixth century, a similar plague destroyed what was left of the Holy Roman Empire, killing perhaps 100 million people and ushering in the Dark Ages. At the beginning of this century, a third great pandemic killed another 13 million people, mostly in the Orient.

It was amid the horrors of this last plague that science finally unlocked the secret to its cause. In 1898 a Japanese researcher named Ogata at last unmasked humankind's most deadly foe—the oriental rat flea. This devious assassin carries the bacterium *Pasteurella pestis* and transmits it from one host to another through its feeding. Bit by bit the mystery was unravelled: the plague bacilli exist in certain rodents, particularly marmots in central and eastern Asia, without much consequence. But periodically the marmots come into contact with rat packs, and fleas hopping from marmot to rat carry the deadly bacilli with them. While feeding upon an infected host, the flea sucks up bacilli. During a subsequent meal on a new host, it regurgitates infected blood into the new host's system. Enough infected fleas transferring onto sufficient roving rats can trigger a deadly infestation.

Rats themselves are not immune to the bacillus, and when a host rat begins to die, the fleas infesting it abandon it in search of another warm-blooded host. Where rats are dying in numbers, humans become a likely target for the voracious fleas. The bacilli incubate in humans for a week or more before symptoms become evident. The telltale symptom is a bubo—an inflamed swelling in glandular parts of the body, especially the groin or armpits. *Bubo* derives from the Greek word for groin, and *bubonic* means "attended with the appearance of buboes."

Fever and delirium ensue and, in about 80 per cent of untreated cases, death. Since the great Indian plague of 1896-1917, fatalities have decreased through the widespread use of antibiotics. But in 1991, the official *People's Daily* reported that bubonic plague is on the rise again in China. Between 1986 and 1990, there were 108 reported cases of the plague in China, more than double the total for the preceding five years. Brendan Lehane writes that bubonic plague has probably caused more human grief and fear than any other single cause, and all the savage wars of human history together, "are as a puff of smoke against the typhoon blast of fleas' ravages through the ages."

* * * * *

The discovery that fleas are the main vector in transmission of the plague triggered a period of intense scientific scrutiny of the biology and ecology of fleas. Over four hundred species were recognized and classified. The largest known flea is a North American giant as long as a pencil-width and described from a single specimen taken from the nest of a mountain beaver near Puyallup, Washington, in 1969.

Dog and cat fleas are among the most numerous, successful and widespread of their race. Neither confines its attentions to its namesake, and each can be found on a variety of hosts, including humans. The common flea (*Pulex irritans*) infests other animals besides humans, but is thought to be gradually losing ground to more versatile dog and cat fleas. Lehane reports that analysis of two thousand fleas collected off humans in Britain determined that over one-half were dog fleas (*Ctenocephalides canis*).

Believed descended from a winged insect related to the common fly, the flea is splendidly constructed for the lifestyle it has evolved. Its body resembles a miniature prawn—scaly, narrow, and streamlined for moving quickly through dense jungles of fur. Its hard head is pointed or rounded. Some species have eyes, others do not, and none relies greatly upon sight. Far more refined is the creature's acute sense of taste/smell. It has a pair of antennae as well as a set of palps near the front of its head. Together these and other sense organs can detect minute differences in aroma, warmth, light, vibrations, etc.

Once settled on a suitable host, the tiny parasite can feed to its heart's content through a complex drill and syphon apparatus. With its clawed feet grasping nearby hairs, it clears away obstructing hairs or feathers, scrapes the skin surface raw with a pointed stiletto, raises its tail and plunges its proboscis into the flesh. Through the proboscis, formed of three stilettos, it injects an anti-coagulant saliva into the puncture and syphons up the free-flowing blood.

Behind its head, the flea's thorax divides into three segments, each with a pair of legs attached, the two marvellous jumping legs being the largest and farthest back. The thorax attaches to an abdomen encased in smooth, hard plates divided into ten overlapping segments. These bristle-covered armoured plates house the circulatory, respiratory, digestive and reproductive organs. The flea can curl into an armoured ball for protection, and it's the plated abdomen which cracks when a flea is given the traditional thumbnail squeeze.

Genitalia are confined to the last two segments of the abdomen, and flea copulation has been hailed as one of the wonders of the insect world. The male, normally much smaller than his mate, slides beneath her from behind, embraces her back-to-belly with his antennae and softly caresses her genitalia. Then his tail curls up like a scorpion's and he penetrates her with what Brendan Lehane calls, "the most elaborate genital armature yet known." The male, he writes, "possesses two penis rods, curled together like embracing snakes. Inside his body, the smaller rod moves outwards lambently, catching delicate skeins of sperm and moving it into a groove on the larger longer rod. Then the whole phallic coil slides out from his sensitive rear, the large rod enters the female and guides the thinner along beside it." The thin rod continues inwards, eventually depositing its sperm and withdrawing. "Any engineer looking objectively at such a fantastically impractical apparatus would bet heavily against its operational success," writes entomologist Miriam Rothschild. "The astonishing fact is that it works."

Mating fleas may remain locked in this improbable embrace for several hours, meanwhile feeding away. Fertilized eggs are laid within a few hours of copulation, and the mating pair may repeat the whole operation. Throughout the mating ritual, the pair continue eating and defecating at an accelerated rate in order to provide food for their offspring. Within a few days, larval maggots hatch from the pearl-white eggs. The maggot spends several weeks feeding voraciously on its parents' excrement, and on dried blood and bits of dead skin, before spinning itself a silky cocoon in which to pupate. Camouflaged by dust and dirt sticking to the cocoon's adhesive surface, the pupating flea may remain in the cocoon for little more than a week, or for as long as a year. Eventually it stirs, responding to movement, and emerges to begin its own adult life of hopping, eating and mating. On average, a flea can expect to live several months, but in exceptional cases may survive a year or more.

* * * * *

There's an oft-repeated old folk tale about a flea-bitten fox who, to rid itself of its tormentors, backed slowly into a running stream, while holding a clump of moss in its mouth. As the fox became progressively submerged, the fleas moved up its body, and eventually out along its muzzle. Finally, with only the tip of the fox's nose exposed, the fleas hopped onto the moss. Once they were all aboard, the fox released the moss and watched it drift downstream, bearing off the outfoxed fleas.

For centuries humans too have tried to find an effective way of getting fleas off themselves and keeping them off. Some of the reported remedies are highly suspect. For example, the Roman writer Pliny advises that if, upon hearing the first cuckoo cry of spring, you collect the earth under your right foot and sprinkle it in your bed and around your house, you will not be bothered by fleas all year. Equally questionable is the suggestion in an ancient Greek treatise on agriculture that a person in a flea-infested locality has merely to cry "Ouch! Ouch!" and the fleas won't bite.

More promising, perhaps, though far from egalitarian, was the sacrificial lamb approach, which involved placing a particularly succulent body in such a way as to attract the fleas to it rather than to yourself. Several sources report that a sheep was often used for this purpose in the southern U.S. An ancient Egyptian manuscript recommended that a person—presumably a slave—be anointed with the milk of an ass and made to stand in the room to attract fleas.

In certain parts of England, tradition dictated that you air the bedding before Easter in order to keep fleas in check. The folklore of other counties recommended that all windows be kept closed and the doorstep be swept on March 1 to ensure a flea-free season. Although fleas don't normally become a problem until the warm months of summer, the month of March figures more prominently in flea-lore than any other. If some fleas managed to sneak into one's house despite March 1 precautions, they could be driven back out by burning a dirty dishcloth upon hearing the first thunder of March. "Kill one flea in March," advises a wise old proverb, "and you kill a hundred." Later in the season, one could drive fleas off by melting in the fireplace some snow which has fallen in May. For many rustic peoples, leaping over the fires kindled on Midsummer's Eve cleansed the jumper of both sins and fleas.

Traditional flea repellents have included everything from cow dung to the splinters of a tree struck by lightning. Dwarf elder leaves and alder leaves have been credited with repelling fleas, as have Chinaberry, spearmint and foxglove, wormwood, pennyroyal and rue. "Above all," advised the Elizabethan writer Thomas Moufet, "the dregs of Mares' piss, or sea-water are commended, if they be sprinkled up and down." Because fleas are drawn to warmth, Lehane reports, some African villagers place a lighted candle in a shallow dish of water in the middle of the room. Hopping towards the warmth, fleas land in the water and drown.

Having "a flea in one's ear" became a popular expression meaning to suffer a stinging or mortifying rebuff or reproof. But in the first century the Roman medical writer Cornelius Celsus addressed himself to the literal problem of extricating real fleas from human ears. In simple cases, he advised, a bit of wool pushed into the ear would entangle the pest and allow it to be extricated. For intractable cases he recommended the following procedure:

> . . . a plank may be arranged, having its middle supported and the ends unsupported. Upon this the patient is tied down with the affected ear downwards, so that the ear projects beyond the end of the plank. Then the end of the plank at the patient's feet is struck with a mallet, and the ear being so jarred what is within drops out.

In our own enlightened Age of the Chemical Fix, we've used DDT extensively against fleas, with disastrous consequences for many creatures. Some pet owners still purchase flea "tags" that hang from pet collars, releasing toxic fumes. Chemical flea collars work by having a dog or cat absorb flea-killer (pulicide) into its system until it becomes too toxic for fleas to survive. Richard and Susan Pitcairn, authors of *Natural Health for Dogs and Cats*, warn that, "some [chemical] flea collars are so potent these days that they produce extreme skin irritation and permanent hair loss in some animals, particularly if they're too snug." Similar warnings apply to the toxic substances contained in many flea shampoos, soaps and powders.

In place of the "make your pet so toxic even fleas won't bite it" approach, progressive veterinarians recommend an improved pet lifestyle: careful grooming, lots of exercise and fresh air, and a healthy diet, along with meticulous house-cleaning. Frequent bathing of pets is one of the most effective flea-control measures. Just regular soap or shampoo will control fleas if a dog is bathed every two weeks or a cat once a month. Some people maintain that putting fresh garlic and/or yeast in their pet's food keeps fleas away, and I do this with our dog in the summer. Other pet owners are equally insistent it's worthless.

If fleas persist, the Pitcairns recommend adding a cup of insect-repellent herbal oil, such as pennyroyal or eucalyptus, to the bath. Frequent grooming with a flea comb and daily sponging with a lemon rinse, made from a thinly sliced lemon steeped overnight in water, help as well. Besides nontoxic citrus oil sprays, many health food outlets and pet shops also stock low toxicity pet shampoos, flea collars and powders. These generally contain natural substances such as pyrethrum, diatomaceous earth, pennyroyal, eucalyptus, citronella and others.

Everyone agrees that treating a pet without treating its house and bedding is a waste of time—new fleas simply hop aboard when conditions improve.

Rigorous vacuuming of furniture, rugs and pet bedding removes flea eggs and larvae, but the vacuum bag must be properly disposed of. In the old days, housewives would coat their wooden floors with beeswax to prevent fleas from laying eggs in the cracks. Nowadays some people dust cracks and crevices with diatomaceous earth, the microscopic blades of which slice through the fleas' protective coating.

Despite the best efforts of housewives and herbalists, exterminators and high-tech scientists, fleas have continued to feast off humans and domesticated animals since before recorded history. Of course, they've suffered a bad image spill with the plague business, and it's difficult now to view them with affection. Still, over the centuries they've played a useful part in puncturing human arrogance. When theologians and philosophers were making a living proclaiming humanity the pinnacle of all creation, cheeky poets could retort that the flea, in feeding off humans, has a more proper claim to be at the apex. So let's leave the flea, disgraced but defiant, on the clever lines of Jonathan Swift from *Poetry; a Rhapsody.*

> Hobbes clearly proves that every creature
> Lives in a state of war by nature.
> So naturalists observe, a flea
> Has smaller fleas that on him prey;
> And these have smaller still to bite 'em,
> And so proceed ad infinitum.

Chapter Four

DEER

The Terrible Beauties

W hen the lashing rains and drifting mists of winter begin to lift in March, and the grass around our house starts to glisten with spring growth, our neighbourhood deer emerge. Sitting at my desk each morning around sunrise, I watch them tiptoe out of the nearby woods, alert and tentative, small bands of nervous survivors from the forest. More slender even than their normal slender selves, they're a disheveled-looking bunch this time of year. Their mousey-brown winter coats are beginning to unravel in spring moult, tattier by the day, like hand-me-down thrift store furs.

They seem all bones and sinew after the rigours of winter, the sudden, explosive terrors of hunting season, and the frenzy of the rut. We've seen very little of them through the dark months when they find nothing much of interest in our little clearing. We know they're still around in the winter— we see their tracks in the wet snow and sometimes hear their telltale thumps as they bound away in the dark. Once in a while, working in the winter woods at getting next year's firewood or at clearing trails, I'll surprise a lone deer—or it will startle me as it bounds off suddenly through the underbrush. But by and large they're not much of a presence.

By March they emerge, bone-thin from winter rations, and crop our fresh lawn grasses like greedy sheep. As the season progresses, they'll try their damnedest to eat everything else we've grown as well; they'll infuriate us, drive us to murderous rage with their cunning raids on our vegetables, fruits and flowers. But in springtime, when all the earth seems young again, when all things seem bright and beautiful and possible, these bony bandits are a welcome sight, as much a part of spring rituals as the returning swallows.

They soon grow accustomed to us as we pass back and forth through the gardens and orchard. They'll stand within ten metres, watching carefully, scenting the air, their sharp ears vibrant for any hint of danger. Then, one day we realize that we haven't seen many of them for a bit. We watch for them at dawn and dusk, but only the odd small buck is to be seen; the rest are back to their old secretive ways again.

The breeding does have simply retreated deep into the forest to give birth, and their small extended families have gone with them. In the mythology of the Kwagulth peoples of this coast, the deer is possessed of supernatural gifts by which it eludes those who would pursue it. One of its tricks is a "fog box" which, when opened by the deer, emits a cloud of fog into which the elusive animal disappears, leaving the hunters confounded. The sudden springtime disappearance of the deer has the same feeling to it—for weeks they're around, as common as stones, then magically they're gone.

By late May we begin to chafe, wondering if there'll be any fawns at all this year. And then in the bright mornings when the woods echo with bird-song, deer begin to appear regularly again. The yearlings and young does step into the clearing gingerly. Gone are their former shabby selves, transformed into everything chic and elegant. They've thrown off their thread-bare winter pelage and replaced it with sleek, cinnamon-coloured summer coats. Their starlet-slender thighs and throats are white, the hair under their legs and belly soft and silky white. Each face is as delicate and lovely as the face of a beautiful youth. The muzzle glistens, black and moist; the pointed ears are outlined in black. Soft and limpid, their large brown eyes are accented by stylishly long dark lashes. Beneath each eye a tiny gland is lined with fine white skin—the so-called "tear pit" whose secretions led tragic poets to believe that deer weep tears of emotion. Thus Shakespeare writes in *As You Like It*:

> A poor sequester'd stag . . .
> Did come to languish . . . and the big round tears
> Cours'd one another down his innocent nose
> In piteous chase.

Timid and shy all over again, they linger at the edge of our clearing, nibbling at huckleberry or salmonberry leaves, or the unfurling tips of sword fern fronds. Finally, marvellously, out from the woods steps a doe with her fawns, and no matter how many years you've seen it, the sight remains a lovely and soul-lifting experience. There are usually twin fawns—tiny and delicate creatures, their soft reddish coats speckled with white spots. On slender legs and dainty little hooves, the fawns seem, like children, the essence of everything sweet and innocent and unsullied in an often too-vulgar world. The dam's caution is redoubled. She lingers near the edge of the clearing, close to safety, keeping her little ones nearby. They suckle at her teats, their silly little tails flicking with pleasure. Here's a vision to banish cynicism, to put loutishness to rout.

Or is it? Walt Disney shrewdly made millions from *Bambi*. More recently, the B.C. tourism ministry set about creating an ad campaign to lure foreign travellers to "Super, Natural British Columbia." The clever folks at the advertising agency decided to create illusions of innocence and wild beauty with pictures of a doe and fawn in a mist-shrouded old-growth forest. In reality, news reports confirmed, the bewildered animals came from a petting zoo, the scene was lit by artificial lighting, the drifting mists poured from smoke sprays and the virgin wilderness was a roadside park. Super Tacky, sniffed the critics.

* * * * *

I prefer the real thing. The wild deer in our yard are Columbian black-tailed deer, a subspecies of mule deer. A creature of foothills and mountains, most at home in open forests, parklands and hillsides, the mule deer roams across western North America from Mexico to the Northwest Territories. Closely related to mule deer, and similar in size and general appearance, is the white-tailed deer, which ranges right across the continent and is probably the most familiar big-game species in North America. The two species are distinguished by certain characteristics: mule deer have larger ears and high, branched antlers. Whitetails have low and compact antlers and a large tail which is always white underneath.

The Columbian blacktail, or coast deer, is confined to the islands of the Pacific Northwest and the western slopes of the Coast Ranges. As its name indicates, blacktails are recognized by distinctive black markings on their tails. Biologists also know them through particular skull, teeth and antler features. On small islands such as the one where I live, the deer are semi-dwarf, seldom attaining comparable size to their relatives on nearby Vancouver Island. Blacktails are a small race generally, and the size of various mule deer subspecies varies considerably. Adult males may weigh anything from 50 to 215 kilograms; females weigh between 30 and 70 kilograms. Whitetails, which are generally thought of as the more graceful of the two species, vary far less in size. Adult bucks weigh between 85 and 96 kilograms, and adult does from 57 to 62 kilograms.

These common North American species are about mid-range when it comes to deer sizes. The smallest true deer in the world is the northern pudu of Ecuador and Columbia. Adults are only about thirty-five centimetres at the shoulder and weigh no more than eight kilograms. By contrast, in Neolithic times, a giant fallow deer was common in Europe, and particularly Ireland. Fossils of these giants, unearthed from Irish peat bogs, indicate an antler spread of almost four metres from tip to tip. Today's fallow deer, the semi-tame species seen in European parks, are smaller than North American whitetails, as are the roe deer common across Europe. Much bigger, and highly prized as a big-game species in Europe, the red deer can weigh up to three hundred kilograms and stand about two metres at the shoulder.

Newborn blacktail fawns weigh between two and three kilograms. Normally they're born as twins, but they may arrive singly or as triplets. The white spots that dot their chestnut coats are formed by longer, coarser hairs that are tipped with white. As summer passes, the hairs wear down and the white tips fade away. In dappled sunlight, the spots create a perfect camouflage, and are a vital part of the fawn's early survival gear.

Normally a dam will cache her babies—separately if she has more than one—while she browses nearby. Instinct instructs the fawn to lie perfectly still, flattened to the ground and silent. Its spots blend into the sun-dappled foliage where it lies, and its mother's incessant licking has removed any whiff of scent. I got a first-hand demonstration of how effective this camouflage can be while hiking one June in the foothills of Alberta, where I almost stepped upon a tiny fawn cached amid straw-coloured grasses.

This motionless and scentless disguise, effective against predators who hunt by scent and sight, is entirely useless in fending off well-intentioned humans. Every spring our local papers carry messages from wildlife officers warning people against picking up what they take to be "abandoned" fawns. And every spring the warnings are closely followed by a parade of Good Samaritans carrying "abandoned" fawns to wildlife shelters. Most of these are fawns cached by the dam, which has then moved off for safety; seldom in nature is a fawn abandoned by its mother. But once it has been interfered with by humans, a fawn's chances of successfully returning to the wild are infinitesimal.

"Man is deer's worst enemy," writes Vancouver Island wildlife author and artist Joan Ward Harris. In her 1979 book, *Creature Comforts*, Harris narrates her experiences raising, among other wild orphans, a young fawn. People think they can feed a fawn milk with sugar added, writes Harris, and in doing so condemn it to death by starvation. "Doe's milk contains twice the solids of Jersey milk—which is the richest cow's milk; more than twice as much fat, three times as much protein and twice as much ash (minerals), but less sugar." For the first few days, she adds, the colostrum-rich milk contains abundant vitamins and antibodies.

In the wild, suckling on this powerful formula, fawns grow quickly from tiny, elfin waifs into voracious brown bundles of energy, bounding about and playing recklessly. After about a month, the does no longer cache them, and this is when they begin to appear in our yard. They're usually weaned at about four months, though occasionally one will continue suckling until its mother mates again. Fawns remain part of the matriarchal grouping throughout their first year, and sometimes thereafter. Not a particularly gregarious species, blacktails seem to prefer roaming in small groups; around our place they usually form bands of three or four. Even as yearlings, most bucks are already solitary.

On the coast, blacktails wander about on surprisingly small home territories; some will spend their entire lives on a single hillside. Mule deer generally are wanderers, many roaming up into alpine pastures in summertime and back down to valley bottoms for overwintering. Whitetails are usually more sedentary and more solitary, and may establish a home range of perhaps two square kilometres.

Deer browse on all kinds of foliage. Blacktails love the new growth of Douglas fir seedlings, much to the dismay of beleaguered foresters. Even western red cedar, which few other creatures can tolerate, is a favourite winter food; so much so that the deer introduced onto Haida Gwaii (Queen Charlotte Islands) have been branded an "ecological disaster" for wiping out cedar seedlings. One particularly bone-headed minister of forests went so far as to blame deer for causing a succession of large landslides that occurred on island mountainsides which had been clearcut!

Less catastrophically, I've watched blacktails browse on blackberry vines, huckleberry and salmonberry bushes, salal and serviceberry, willow and dogwood leaves. On the mainland they add bitterbrush, mountain juniper and sagebrush to their diet. Around our place they favour grasses and clover in the springtime, later shifting over to shrubs and vines. By winter they've retreated into the forest where they browse on understorey vegetation, and particularly on green-grey beard lichens that have fallen from tree branches.

All this assorted foliage disappears into a large cavity called the rumen, where a host of microorganisms and mild digestive juices begin breaking it down. Like other ruminants, a deer can "chew its cud" by regurgitating the partly digested greens from its rumen. Eventually the predigested curds pass on to the true stomach, or abomasum.

* * * * *

Most of all, it seems, deer like to eat whatever humans have planted for their own use, and preventing them from doing so requires tremendous tactical skill, ingenuity and vigilance. Safe from wild predators and hunters, deer populations have increased dramatically in some suburban areas, and humans have encroached more and more onto traditional deer ranges, so that deer are now considered a serious pest in some regions.

There's a fabulous body of folklore and fiction around how to defeat garden-loving deer, and much of it is utter rubbish. One of our otherwise-useful gardening books, for example, seduces the gullible with a list of "deer-proof plants." Some of them I agree with whole-heartedly; experience verifies that deer refuse to browse on daffodils, foxgloves and oriental poppies. But tulips? In our country you could no more get an unprotected tulip to bloom than you could get it to hum Mozart. The same book rates rhododendrons as one of the best "deer-proof plants," and some rhodos do survive handsomely. But we've also had rhodos torn to pieces by hungry deer. Timing often determines what is eaten. For example, our exposed English ivy goes untouched throughout the growing season; but come January, after killing frosts have blackened most other leaves, deer will sneak in under cover of darkness and totally strip the ivy vines. "I don't think there's anything they won't eat if they're hungry enough," one disgusted gardener told me after a nighttime raid on her shrubbery.

It's far easier to identify plants that deer crave than those they don't: just turn your back for five minutes. They'll go to extraordinary lengths to gobble brassicas—broccoli, cabbage, cauliflower and others. Even exotics like spiderwort or Peruvian lilies get munched right down to the ground. But I think roses are their very first love, and when our security system has been temporarily breached, it's always the roses that are the initial victims.

English writers Maureen and Bridget Boland, in their slightly and delightfully eccentric book *Old Wives' Lore for Gardeners*, proposed a rose defense strategy based on outgrowing the deer.

After we had lived here a short while we realized that we could never sacrifice the sight of the deer, at sunrise and dusk, passing through the garden and pausing to drink at stream or pond, but all the young shoots of our roses were nibbled off. We planted enormous tree-climbing varieties like Himalyan musk and Kifsgate which grow to thirty or forty feet, and protected their lower stems with chicken wire while they were young, and such huge shrub roses as Nevada, whose lower, outside shoots alone suffered.

As their battle of wits continued, the Bolands write, "We read that sprinkling lion manure would terrify the deer, and could well believe it." Indeed, there's a whole school of thought dedicated to the use of savage excrement as deer repellent. In these circles, tiger dung is considered an adequate substitute for lion excrement. The indefatigably organic editors at Rodale Press in Pennsylvania report that one gardener protected his apple trees with truckloads of manure from a nearby zoo, hanging bear, lion and tiger droppings in small bags throughout the orchard.

Where exotic excrement is in short supply, we're advised by organic gardener Beatrice Trum Hunter, "a few pails of human urine, placed at the corners of the garden, as well as in the middle, deter deer." And everyone else too, I should think. Dog's hair, as well as human hair from barbershops, sprinkled near plants, are also credited with keeping deer at bay. A devout skeptic in such matters, I've never tried either type of hair—but I have watched deer browsing contentedly within seven metres of our lazy old border collie, Fen, asleep on the lawn. Somehow I think that if the scent of a living dog doesn't do the trick, there's not much hope for a sprinkling of dried-up old dog hair.

Proponents of these offensive odor strategies seem to keep testing new substances with all the fervor of cold warriors developing new weapons systems. Bars of bath soap hung on ropes from the branches of fruit trees, dried blood meal fertilizer hung in old nylon stockings, moth balls in net bags: all have been promoted as odd-looking but effective repellents. The abiding problem with most of these foul-smelling solutions is that they leach away in rain and sun and need to be constantly replaced.

The deer- and rose-loving Bolands investigated these avenues and were rewarded when "an Old Wife provided a much easier solution. Tie an old piece of thick cloth such as flannel on the end of a bamboo cane and dunk it in creosote, and stick it in the ground like a little flag near each rose, or at the corner of a bed. The deer will not risk coming near the strong smell, which will prevent them from scenting the approach of danger." For purists who would object to sniffing a toxic and carcinogenic rag when they go to smell a rose, the Bolands hasten to add, "after a day or so the smell will not be apparent to humans unless they actually sniff the cloth."

An equally offensive chemical fix is provided by the pesticide Thiram, used, my pesticide applicator's handbook tells me, "in a variety of formulations on fruit trees and woody ornamentals to repel deer, mice and rabbits." This charming substance, the handbook warns, can cause eye, nose, throat and skin irritation. "Launder contaminated clothing before reuse," it cautions, and "avoid alcohol consumption prior to and after working with Thiram."

No, thanks. I'm a fence man myself, and a fence man I'll remain. Our vegetable patch is surrounded by a stout two-metre-high barricade made of horizontal split-cedar rails interlocked and with gaps between them of no more than eight centimetres. It has proven entirely effective for many years—in part, I believe, because deer won't voluntarily leap an obstacle unless they can clearly see where they'll land on the other side. The blacktail is a mighty leaper; it's known in places as "jumping deer" from its peculiar, stiff-legged, bounding gait when frightened. We have a small advantage here in that our local deer are semi-dwarf; gardeners on Vancouver Island and the mainland need at least 2.5-metre fences to keep deer from jumping over. I've read of one Oregon gardener who got away with a 1.5-metre-high chicken-wire fence with another single strand of wire running along a metre outside of it and about a metre above ground level.

Much like the Bolands, we didn't want to deprive ourselves of the charming sight of deer browsing on the lawns beyond our house. We used traditional split-cedar rails in areas where visibility wasn't a consideration, and where we wanted an unimpeded viewscape, we designed a rose arbour constructed of 2.5-metre-high cedar posts connected on top with thin horizontal rails. Galvanized stucco wire, 1.5 metres high, strung between the posts, keeps the deer out—they won't try to leap between it and the rails above it— while preserving a sense of openness to the lawns beyond

The system is almost perfect—but leave a gate open by oversight, and the deer will discover it before you do. Last year two little fawns became regular intruders; they were small enough to squeeze through the gaps in the split rails. Once they'd squeezed through and begun feasting on our roses, their mother was frantic to join them. She'd try anything, including climbing the cedar rails, and knocking half of them down in her clumsy, splay-legged attempts. Mercifully, the fawns soon outgrew their entry gaps and the problem disappeared. I believe that in defensive strategies against these persistent marauders, "almost perfect" is as close to perfection as you'll ever get.

* * * * *

As the days shorten and howling southeasters begin to blow, the deer moult again, putting on a thick, coarse grey-brown winter coat. The pelage sheds water admirably, allowing deer to stand in our yard contentedly feeding amid drenching downpours. Snow is another matter, and in mountainous country, it drives the deer down to sheltered winter ranges in the valley bottoms.

By mid-October the bucks are beginning to rut. By then they've rubbed their small antlers free of summer velvet and take to polishing them on shrubs and saplings as a form of erotic stimulation. They become surly and truculent. Each will defend his territory against other bucks, shoving them off with his antlers. Docile through the rest of the year, a buck in rut is a formidable opponent. He lowers his head and presents his antlers like duelling weapons. His upper lip curls almost to a lewd sneer. He bunches the muscles on his hunched back, and hair stands along his spine like hackles on a fierce dog. His tail is upright. He stamps and paws the ground impatiently and utters throaty, threatening coughs. More than one amateur wildlife photographer, thrilled that a deer was staying put for a real close-up shot, has had a rough introduction to the violence of the rut with 150 kilograms of angry and antlered muscle in pursuit.

As the does come into heat, the "running of the deer" begins. With a receptive time slot no longer than a day, the doe runs through the woods in search of a buck, marking her trail with urine and scent gland secretions. The polygamous buck will follow one doe for a few days, then abandon her for another. If unsuccessful in mating, the doe will enter heat again in about four weeks, and breeding may extend until early March, when the exhausted bucks shed their antlers and return to their old docile ways. For the does, who have reached breeding age in their second or third year, the gestation period is about seven months.

* * * * *

Seldom do I go tramping through our little island's woodlands without finding at least one deer carcass. Some of these, no doubt, are the handiwork of unskilled hunters who have wounded but not killed their prey. Up until recently it was common practice for some islanders to go "pit lamping" for their winter meat—mesmerizing roadside deer at night by shining their headlights at them and then blasting away from inside their pickups. Nowadays the great white hunters at least pursue their quarry by daylight and on foot, though the sound of their shooting still gives me shudders. Brigitte Bardot once described the annual French deer hunt as "a cruel bloody sport practised by a stupid pretentious elite."

In places of harsh winters and deep snows, winter starvation is a major cause of death in deer herds. Worse for our deer are the predations of domestic dogs running in packs. It's a growing problem on the islands, and in valley bottoms where increasing numbers of dog-loving humans are usurping deer winter range. Wildlife officers report that a deer pursued by dogs will often die a particularly gruesome death. The dogs will chase it for sport, not trying for a quick and clean kill the way a hunting predator would. Instead they'll tear at the deer's flanks and run it until it drops from exhaustion or is cornered by a fence.

More natural predators hereabouts are wolves and cougars. Bear, lynx, coyote and bobcat also hunt them, and it's said that golden eagles will sometimes manage to snatch a fawn. Parasites and diseases take their toll as well. Our browsing deer are constantly flicking their tails and twitching nervously all over from skin parasites—ticks and deer louse flies—deer flies, blackflies and other tormentors. Blacktail deer have been known to live as long as twenty years, but biologists say a twelve-year life-span is more typical and that, with hazards at every turn, many don't live even as long as that.

* * * * *

Several years ago I found myself perched high on a mountainside in the Nimpkish Valley on northern Vancouver Island. In the late sixties and early seventies the Nimpkish was renowned for its deer herds and known as "the home of the deer." Tales were told of logging crews who'd start work a half-hour early just so they'd have time to shoo fawns off the logging roads. Opened to large-scale clearcutting thirty years ago, and only made accessible by road in 1965, the two-thousand-square-kilometre valley was ideal deer habitat: vast stands of old-growth Douglas-fir provided winter range during heavy snowfalls, and new clearcuts opened up abundant summer browse.

As I stood at dawn on the mountainside overlooking the valley, I could trace through the lifting mists a network of logging roads crisscrossing the hillsides. A million cubic metres of prime old-growth timber is taken from the valley every year, and some of the clearcuts cover more than two thousand hectares. But the wildlife biologist with me, scanning the valley with binoculars, was unable to sight a single deer. Ironically, there are more of them in my yard than in this vast area which was once renowned for them.

What happened? In the early 1970s wolves reappeared in the watershed—lots of wolves. Where they came from and why they came just then nobody is certain. But they were quickly targeted for causing an 80 per cent decline in the once-legendary Nimpkish deer herds. The province's Fish and Wildlife Branch, committed to providing hunters a minimum harvest of ten thousand deer per year from Vancouver Island, came under intense pressure to eliminate wolves. Hunters took to indiscriminate shooting of wolves. Other geniuses began scattering strychnine and cyanide-laced baits.

The debate over wolves, deer and old-growth trees became increasingly strident and increasingly complex. As more and more of the ancient forest is cut down and trucked away, the deer's vital winter ranges shrink. Conservationists argue that forcing the deer to hole up in smaller and smaller pockets of old-growth is simply making it easier for wolves to find and kill them. The timber company argues that it's unrealistic to leave large patches of old forest, at a cost of millions of dollars and hundreds of jobs, just so deer can congregate there and be slaughtered by wolves. The hunter lobby presses to have wolves poisoned so that it can claim its share of deer. Self-appointed vigilantes attack and harass the wolf packs. And provincial politicians—who must ultimately make the difficult decisions about preserving old-growth timber and poisoning wolves—continue to dither. The Nimpkish imbroglio, in what was until three decades ago a magnificent and unspoiled valley, speaks volumes of how little we have learned over the centuries—or perhaps how much we have forgotten—about ourselves, our fellow creatures and our rightful place in the natural world.

Chapter Five

SLUGS

Nature's Slimy Recyclers

S lugs. The very name excites images of things loathsome and repugnant. Arguably the most despised creatures in creation, they are cold-blooded, slippery, slow and, most awfully, slimy. They slither about our gardens under cover of darkness, tearing ragged holes in the salad greens, gnawing filthy chambers in potatoes, their excrement defiling the curds of cauliflowers. We call things sluggish, meaning slow and unresponsive. "You slug!" we taunt our enemies. "You slimeball!"

When I first glimpsed British Columbia's legendary "banana slug," I was freshly arrived on the West Coast from Toronto, and nothing in the East had prepared me for this obscenity. The slug looked to be about thirty centimetres long—a sickly, jaundiced yellow, splotched with black. It oozed vast quantities of slime, leaving a silver ribbon where it passed. Its tentacled head swung slowly from side to side, absurdly menacing. An involuntary shiver rippled down my central nervous system, as my stomach churned. I had met the nation's biggest slug but little understood how much it had to teach me.

When I eventually settled on the coast and became something of a gardener, my relationship with slugs both intensified and took a decided turn for the worse. Whole spinach plants would disappear overnight. The succulent leaves of hostas would be cut to dirty ribbons. Slime trails would wind suicidally along the thorny arms of rose bushes, their leaves now mutilated. Tulip tips would be snipped like cigars. When marauding slugs finally invaded the greenhouse and set upon my precious eggplants, I declared war.

It was a just war, I told myself, a holy war. I spoke the implacable language of the cold warrior: self-defence and deterrence. After every rain I prowled our garden pathways armed with a stick sharpened to a deadly point. I'd spot a slug, poise the murderous weapon above its mantle, then fiercely drive it down. Often I'd spear two or three dozen big banana slugs in a single expedition. I began to take pride in my kill ratios. I started to boast. Ernest Hemingway among the slugs.

Then I began to realize that these were forest slugs; I saw them in the dark conifer woods all about. Undoubtedly they've been here since the last ice age, part of the forest ecology, so who am I to slaughter them for my own purposes? Glancing through a magazine one day, I was alarmed to discover that the banana slug is not much of a garden pest at all. The real damage is worked by a half-dozen other species, all smaller and less easily detected. Whatever smaller slugs I'd seen—like those little white ones in our spinach salads—I'd taken to be babies of the big ones. Colossally ignorant, I'd been impaling benign natives, blaming them for the predations of distant relatives. And, in the ultimate irony, the real culprits were all introduced here by Europeans in the first place!

Chastened, slightly shaken, I set about educating myself. The literature on slugs is not vast, nor do experts on them abound. Gardening books often contain a section on slugs as pests, their tone typified by one description of them as, "a thoroughly nasty bit of business." The most comprehensive work I found was titled *Terrestrial Slugs,* by British scientists N.W. Runham and P.J. Hunter, which details the biology, physiology and ecology of terrestrial slugs in readable English. Various research papers and doctoral theses helped fill out the picture, though several academics told me that much work remains to be done on these strange creatures.

What we do know already is fascinating. Slugs—essentially snails without shells—belong to a large animal phylum called Mollusca, which includes oysters, clams and other shellfish. A large group of slugs, including the sea slugs, are members of a class of molluscs called Gastropoda—a huge and varied category numbering many thousands of species. Narrowing further, terrestrial slugs are lumped into seven main families, but there's much confusion about precise species distribution and identification, due to remarkable variations in colour and size within species.

Uncertainty also clouds the land slug's evolution. Lacking hard body parts, the slug has left no fossil record of its evolutionary path. Once upon a time, all Mollusca inhabited the shallow littoral zones of the sea. But despite their primitive appearance, molluscs are a tremendously adventurous and adaptable group. Over the millenia they've discovered an extraordinary range of environments in which to live. Terrestrial slugs are believed to be descended from sea slugs, which gradually worked their way into fresh water and then onto land. The slug's gradual shucking-off of its snail-like shell is another adaptation that enables it to survive in environments such as highly acidic peat bogs where snails are not successful. And, lacking a shell, the slug is far more streamlined and able to squeeze into small cracks and crevices for shelter.

* * * * *

Slugs like best a cool, moist environment; frigid winters and roasting summers are not their style. So it comes as no surprise that the mild and moist Pacific Northwest is some of the best slug habitat found anywhere. There are about a dozen major species of slugs found here, some of them natives, others European immigrants; many of them are found elsewhere across the continent.

Ariolimax Columbianus, the giant banana slug whose appearance so assaulted my eastern sensibilities, is a native of West Coast temperate rain forests.

It can grow up to thirty centimetres long, and comes in a range of colours—from pale yellow to brown to glistening black. Many are mottled with black spots. Measuring less than five centimetres, the light brown field slug, *Ariolimax laevis* is common right across the country. Another slightly larger native, the greyish brown *Ariolimax reticulatum,* is also widely distributed and is particularly abundant in the Maritimes.

While these and other native species may cause gardeners and farmers some trouble, their predations are nothing compared with the ravages wreaked by a number of imported garden slugs which have multiplied rapidly in the New World. Perhaps the most hated of all is *Arion ater*, a European immigrant known as the black slug, though it also comes in brown, red, green and yellow. It attains a maximum length of about eighteen centimetres and has a ravenous appetite, consuming many times its own weight in food every night. Believed to have been introduced in the 1940s, it is now a prevalent species throughout the Pacific Northwest and causes substantial damage, particularly to strawberry and lettuce crops.

The grey garden slug, *Deroceras reticulatum,* is a smaller, but very common, field and garden pest. The midget milky slug, *Deroceras agrestis,* is another nuisance which gets its name from its particularly sticky, opaque white slime. Also widespread is *Limax maximus,* known as the spotted garden slug or, more grandly, the great slug of Europe. Grey, with a marbled pattern of black spots, it grows up to eighteen centimetres long.

If any slug can be said to live in the fast lane, it is *L. maximus.* It can travel four times as rapidly as the lumbering banana slug. It's also very aggressive and will kill and eat other slugs, even turning to cannibalism of its own species in captivity. Most notably, it performs spectacular courtship and mating rituals, which we shall get to in a moment.

Other slug oddities include *Hemphillia.* Found under rotting logs, this group of small native slugs twitch violently when disturbed. Then there's *Testacella haliotidea,* an imported species that eschews vegetarianism in favour of earthworms. Another slug, *Milax gigates,* specializes in greenhouses, with a particular fondness for marigolds, geraniums and snapdragons. And there are entirely subterranean species, such as *Arion hortensis* and *A. cicumscriptus.*

* * * * *

Like other molluscs, a slug's body has two components: a visceral mass which houses most of the internal organs, and a combined head and foot.

It has no skeletal parts, though some slugs retain a remnant shell in an internal shell sac. The muscular system is surprisingly complex. The foot sole ripples with a thick network of muscles that move the creature along on a series of wave-like contractions. Another set of muscles in the body wall controls postural movements, while a set of internal muscles allows the slug to retract or extend its tentacles and genitals.

Where its ancestors once bore a shell, the slug has a noticeable hump on its back, called the mantle. Below or behind the mantle are three apertures, always on the animal's right side: a breathing hole called the pneumostome, through which air is pumped to a simple lung; the anus, usually farther back; and an undetectable genital opening.

Twin sets of tentacles extend from the head, the longer (optic) pair having a small black eye at each tip. Researchers consider it unlikely that the eye can perceive detailed images, but rather that it is highly sensitive to changes in light intensity. The smaller anterior tentacles assist the animal in smelling, and perhaps tasting, food. The simple mouth at the front of the head contains a sharp rasping strap, called the radula. It is used to tear food into small pieces which are then passed back through a cavity, to be digested.

* * * * *

Nothing so much typifies a slug as its slime, and nowhere is the creature more wrongheadedly reviled than in its slime output. Slugs aren't just slimy— they're connoisseurs of slime. They secrete and employ different kinds of mucus for use in locomotion, self-defence, temperature regulation, and mating. A substance of remarkable viscosity and tenacity, mucus is composed of proteins and complex hydrocarbons.

Just behind the slug's mouth, its pedal gland secretes a particularly thick, sticky mucus. To propel itself forward, the front part of the pedal gland touches the surface to stick the mucus, then as the animal crawls over it the mucus is extended. The slug literally glides along, even over sharp and abrasive surfaces, on a smooth carpet of its own making. A second, more watery mucus is secreted by unicellular glands, and moved by cilia to the edge of the foot. Researchers believe the watery mucus may serve as a lubricant between the foot surface and the stickier pedal mucus.

Slugs are mostly composed of water, and their slime is designed to absorb water, helping prevent dehydration. That's why, if you get slime on your hands, you can't wash it off. The slime just absorbs water and gets even slimier. Instead just rub your hands together; the slime sticks to itself and rolls up into a ball the same way glue does.

Some species, when attacked by beetles or other enemies, suddenly exude large amounts of a thick, defensive slime; and certain slugs have been observed following slime trails to find their way back to shelter. The versatile mucus also plays a significant role in sluggish lovemaking.

Slugs are hermaphrodites, possessing both male and female sex organs. This is thought to be another adaptation to ensure reproductive success, since an individual can fertilize its own eggs. But copulation and mutual fertilization are common. The courting ritual may include lunging, biting, mantle-flapping, and tail-wagging, and sometimes involves one slug assuming a dominant or aggressive role.

The most dramatic mating rituals are those of *Limax maximus,* and I found them splendidly detailed in a 1987 University of British Columbia thesis by PhD candidate David Rollo on the behavioural ecology of terrestrial slugs. In observing eight species of slugs kept in specially designed cages, Rollo frequently saw a slug locate its prospective mate by following a slime trail. With *Limax maximus,* he notes, courtship is initiated when one slug bites another. The victim flees, and the aggressor pursues, apparently tasting the slime trail for clues. The courtship chase might cover several metres, and climaxes with the slugs climbing to the top of a plant or other raised area.

There they circle around one another, touching with their tentacles, drawing ever closer and eating each other's mucus. This might last half an hour, punctuated with occasional bites. Finally, in limaceous climax, they intertwine completely and together fall into space, releasing a string of mucus attached to the support. Then, hanging suspended together like acrobats, they evert their sex organs and engage in a mutual exchange of sperm.

After this wild exhibition, Rollo continues, the lovers sometimes separate peaceably, but often intercourse is followed by aggression. One or both slugs begin biting and, understandably, the sex organs are quickly withdrawn. Usually one mate ascends the mucus string while the other consumes it or drops to the ground. But Rollo observed that aggression continued in some cases, and in extreme circumstances, the victim had its penis chewed off. In *A Natural History of Sex,* Adrian Forsyth speculates about this behaviour.

> *Ariolimax* often attempt and achieve appophalation: that is, one manages to gnaw off the penis of the other. The castrated slug cannot regrow his penis and is now obligated to be a female and forced to offer eggs....
> It may be that the castrator can raise his reproductive success by increasing

locally the density of females. No evidence of this exists but as it is known that slugs are sedentary and territorial, the idea has scope.

* * * * *

Slugs characteristically lay their eggs at a specific time in their yearly cycle—generally spring or fall—but they also have the ability to withhold egg-laying if conditions aren't right. Often, some time elapses between fertilization and the laying of eggs. Some species actually construct a primitive nest, while others are content to simply deposit their eggs in sheltered crevices where they'll be protected from temperature extremes.

Biologist Alan Carter, who has studied the banana slug, says it lays only about twenty to thirty eggs, but that "it's quite a spectacular thing, because they look like little chicken eggs—they're calcerous and hard, which may be an adaptation to living on land." Carter says the big slug's eggs will overwinter in the ground and hatch the following spring. Other species may lay over one thousand eggs in a number of different batches, and often eggs hatch within two or three weeks of being laid.

David Rollo says that most slugs can be classified into three growth phases: a slow-growing infantile stage, followed by a fast-growing juvenile stage and finally a slow-growing adult phase. Alan Carter adds that banana slugs reach sexual maturity after the first year and live four or five years. For many species the life cycle is about half that.

Young slugs disperse from the nest and eventually settle down to something of a "home range." Their settlement density will depend upon available food and, most importantly, shelter. Almost all of them are nocturnal feeders, needing to retreat to shelter before the heat of the day, although certain woodland natives remain active on cloudy, moist days. Runham and Hunter observed that "time-lapse photographic records of the nocturnal excursions of *Ariolimax reticulatus* seemed to show definite tendencies to return to the place of shelter from which they had started." Another study in a Hertfordshire garden reported that "slugs tended to follow a circuitous pattern, bringing the slugs at the end of each excursion into a position near their starting point."

David Rollo's observations led him to conclude that slugs follow general, but not rigid, routines through the day and night. For the most part, he recorded, slugs remained in a resting position during the day, with bodies contracted and tentacles withdrawn. Nighttime activities included crawling, resting and eating, but in bad weather the amount of time spent crawling and resting diminished. In fact, slug behaviour is regulated by a host of environmental factors, including time of day, surface temperature, light intensity, wind speed, moon phase, atmospheric moisture and barometric pressure.

* * * * *

As beleaguered gardeners know, slugs have a taste for many kinds of leaves, stems, bulbs and tubers. They also browse on fungi, lichens, algae and animal faeces. Equipped with an extremely efficient digestive system, ranging up to 90 per cent assimilative capacity, slugs are among nature's most proficient recyclers.

One analysis of the digestive tract and droppings of *Ariolimax reticulatus* found it feeding principally on the leaves of creeping buttercup and stinging nettle. Flesh-loving *Testacella haliotidea* likes to eat earthworms. Runham and Hunter describe how its needlelike teeth, "converge and clamp round the body of the worm," like the steel jaws of a spring trap, the hapless worm being dragged in through the mouth in stages and slowly digested—a process which may take several hours.

Slugs can detect food from a distance, typically raising their heads and manipulating both sets of tentacles. They'll locate and move towards a desired food, and if you change the location of the food while they're en route, they'll reorient and set off in the new direction.

Competition for territory and shelter seem to be the main reasons for aggression. Although some species are more aggressive than others, slug battles—which are always seasonal—can be surprisingly fierce. Rollo described a typical confrontation as beginning with a touching, withdrawing and re-extending of tentacles. Often an attacker will rear and lunge repeatedly; or the aggressor will push its head into the other's body, biting up to one hundred times. Rollo frequently observed aggressive slugs pursuing victims by following slime trails in order to mount another attack. Besides fleeing, victims may engage in tail-wagging, lifting the tail and wagging it vigorously so as to strike the pursuer a blow. Another form of defence is to withdraw tentacles and head under the mantle, flaring its outer edges to create a "vacuum seal" around the head area.

Slug predators include many birds, small mammals, and reptiles. I've often watched a garter snake slipping through our flower beds with a great fat slug firmly impaled on its fangs. Less daintily, domestic ducks will scoop up slugs by the dozen, particularly in the spring. Unfortunately, slime often coats the birds' bills, and their attempts to wash it off in water produce entirely repulsive, sodden and slimy beards. Turtles, hedgehogs and skunks are said to relish slugs, while gulls and starlings following a plough will devour slugs brought to the surface. Runham and Hunter found that slugs are often infested with parasites which do not bother the slugs, but which may be passed on to domestic or wild animals.

Home gardeners besieged by slugs can look first to making the environment less amenable. Since slugs need a cool, moist shelter relatively close to food, a lot of problems can be solved by cleaning up debris, particularly old boards or shingles piled in a shady corner. Less irrigation and frequent hoeing dries soil out, making it less hospitable to slugs. Scratchy materials like cinders or crushed eggshells spread around plants will also discourage them. I've read that diatomaceous earth works, but several friends have said it's useless for slugs.

There are any number of slug poisons available—the packages often show menacing slugs and snails that loom like Darth Vader—but most contain either metaldehyde or carbonate compounds, which can be toxic to earthworms, mammals, pets and children. Instead, a hollowed-out and inverted half-grapefruit rind can be used as an organic trap. Slugs love to eat the white inner rind and can be caught there in the morning. A saucer of stale beer is another old recipe. Half a potato impaled on a stick and pushed underground can be effective in attracting small subterranean slugs, which can then be dispensed with.

* * * * *

A few years ago when the staff of Richmond Nature Park, near Vancouver, were looking around for a creature that was "familiar but misunderstood," they settled upon the slug, and began an annual festival called Slugfest. Held each June at the forty-hectare bog park, Slugfest draws dozens of youngsters eager to enter their pet slugs in contests for the biggest, smallest, loveliest and slimiest slug. The feature event is a race, with all slugs placed around the perimeter of a half-metre-wide circle. First slug to the centre wins, usually crossing the wire in about two minutes. Later the kids are given a pamphlet on "Care of Your Pet Slug."

The kids may have company soon. After years of being held in contempt, slugs are enjoying a public relations renaissance of sorts. Newspapers report that researchers are studying slug slime for clues to the cure of cystic fibrosis. The communications giant A.T.& T. has featured slugs in an advertising promotion for a new generation of computers. "The slug as savant," trumpets the ad. "Nature has shown us there are powerful computer designs very different from conventional machines." There's an accomplished woodworker in Washington who now makes his living turning out wooden slugs designed to hook over the lip of your wine glass, as though taking a sip. I've also been given a wax slug, a ceramic slug and a chocolate "Slug Sucker" on a stick! Suddenly, it seems, slugs are chic.

Early one morning I spotted a giant banana slug slithering across our garden path. For me, newly educated, curiosity had entirely replaced contempt. I picked it up.

After a few moments of frightened withdrawal, its head emerged, the slender optic tentacles extending like time-lapse flowers. As it began to sense the world around it, one—then both—of its smaller tentacles touched the surface of my skin delicately. I watched the animal apply its pedal gland to my palm. Attaching a silvery ribbon of mucus to my skin, it slipped forward—a cold, smooth softness on my hand.

As I carried it out of the garden, I thought of my past atrocities, of human blundering, of our capacity to wreak havoc on other creatures without knowing anything about them and without realizing what we're doing. Then I released the slug into its native woods nearby.

Chapter Six

DANDELION

Tramp with a Golden Head

*T*o see the common dandelion scuffling about on trashy vacant lots, lurking in unlikely alleyways and sprawling brazenly across wastelands, you'd hardly think that it was once the darling of society. The scion of a fine old family gone to seed—that's dandelion. Wise apothecaries once cultivated it with particular care. Pilgrims and pioneers carried its precious seeds in their sacks, and horticultural clubs fussed over its hybridization. Alas, though its name and lineage are noble, dandelion has run off to become a cosmopolitan weed, a horticultural hobo, a carefree knight of the road.

Dandelion's roots are deep and vaguely Mediterranean. Some will tell you that it stemmed from ancient Greece, and when valiant Theseus had slain the dreaded Minotaur, legend has it that the goddess Hecate rewarded him with a salad of dandelion greens. We know that the Greeks of old tenderized its leaves for salads by blanching, and assume that Hecate did as much for her battle-weary hero. Others claim that Persia is the plant's first home, belonging as it does to the genus *Taraxacum,* a name which the *Oxford Dictionary* says derives from a Persian word. There's debate about this too, and some authorities insist that *Taraxacum* has two Greek words for its root. Dodging protracted debate, others are content with vague allusions to origins in Asia Minor.

Wherever it first developed, dandelion soon succumbed to wanderlust. By their time of empire-building, the Romans had learned that the plant is excellent in both salads and stews. When footsore Roman legionnaires clanked over the Alps into the wildlands of Rhineland and Gaul, who should be there ahead of them but dandelion, nodding brightly along the footpaths and ready to provide a good meal. By the time Caesar got to Britain, he found dandelion already well settled and a staple of the Celtic diet. Typically, the Celts had also learned to brew a heady wine from its flowers.

Dandelion's roaming swept it north into Scandinavia and east to the Orient. The Chinese named it "Nail in the Earth"—a reference to its hearty taproot which, along with its leaves and flowers, they added to their considerable lists of culinary and medicinal ingredients.

* * * * *

Thus spread across the temperate regions of the Old World, dandelion assumed a place of privilege which it retained for millenia. Normans called it *dent de lion,* because they fancied that its toothed leaves resembled lion's teeth. Anglo-Saxon serfs, perhaps with the purposeful obtuseness of the conquered, corrupted the term to "dandylion," and so we have it today. Its correct botanical name, *Taraxacum officinale,* testifies to its use by early pharmacists, for the name implies "a remedy for disorders." In country districts it went by any number of nicknames, some of them downright rude:

blowball, cankerwort, milk witch and monk's head, as well as yellow gowan, Irish daisy, peasant's cloak, wet-the-bed, piss-a-bed and worse.

By whatever name it went, people knew that all parts of it are edible. I first tasted the leaves on a remote peninsula of the Peloponnese. Prepared by a Greek woman in whose orchard we were camping, they were tossed with virgin olive oil and fresh-squeezed lemon juice. When I asked what these wonderful greens were, and she pointed, laughing, to common dandelions growing beneath an olive tree, I felt—well—foolish.

Both roots and leaves were once sold commonly in markets across France. Besides using it as a pot herb, the French enjoyed the leaves and sliced roots on buttered bread. For a gourmet spring dessert, they'd dip the flower heads in batter and deep fry them in oil.

I've not developed much of a relationship with dandelions; I'm not sure why, except, perhaps, that they've only recently invaded our woodland gardens in any numbers. My friend Phyllis remembers picking them as a girl for her Italian mother. She tells me the leaves are best picked before flower heads appear and that they shouldn't be picked in sunshine, or they're apt to taste bitter. In *Edible Wild Plants of North America* authors M.L. Fernald and A.C. Kinsey advise: "Like chicory, the leaves of the dandelion may be blanched by covering during their rapid growth and then prepared as a salad; but the best salad from these plants is prepared from the cold, cooked greens thoroughly chilled, chopped, and served with a proper dressing." Other writers recommend covering plants with an inverted clay flower pot in autumn to produce blanched fresh greens well into winter.

The long, fleshy taproot can be dried, ground up and roasted for a healthy and soothing alternative to coffee. The roots may also be cooked like parsnip, parboiled and then fried. All parts of the plant are useful, and the flowers make a delightful wine, still prized in certain country districts of the Old World. There are many recipes, including one in English writer Lesley Gordon's lovely book, *A Country Herbal.* "Made in May or June," writes Gordon, "the wine will be ready for Christmas." As well, she points out, you can make a good beer from dandelions, and in Victorian times in the north of England, tipplers could buy a bottle of dandelion stout for tuppence.

*   *   *   *   *

"You see here what virtues this common herb hath," wrote Nicholas Culpeper in his *Complete Herbal,* "and that is the reason the French and

Dutch so often eat them in the spring." In the old days people didn't know that dandelion is a rich source of vitamins—it has twenty-five times more vitamin A than tomato juice—as well as of iron, potassium, calcium, magnesium and other healthful elements. But they were intimately familiar with its healing properties, and went to it as readily as we to a drugstore. The Anglo-Saxons used the plant to prevent scurvy and to unblock both bowels and bladder. "So effective as a diuretic is this herb," writes Lesley Gordon, "that it's known as piss-a-bed, pee-a-bed and mess-a-bed in the U.S.A. and Britain and *pis-en-lit* in France." In some English country districts, she reports, children would be given dandelion flowers to smell on May Day to inhibit bed-wetting for the rest of the year.

The plant became a standard medicinal herb in the physic gardens of the great monasteries and of village herbalists. One Restoration diarist referred to it as "a cleansing spring green," and in many places an infusion of its roots or leaves was taken as a blood purifier and spring tonic. Kidney and liver problems, skin diseases, rheumatism, dyspepsia, stiff joints—there seemed no end of ailments that couldn't be treated with a dose of dandelion. Even the wine was credited with strong medicinal properties—always a fine justification to those for whom justifications are required.

The milky juice that exudes from fresh-cut leaves and stems was rubbed on pimples and warts, earning for dandelion yet another nickname: devil's milk plant or devil's milk pail. In Silesia, dandelions gathered on St. John's (Midsummer) Eve were believed to have extraordinary qualities, including the power to repel witches. The Irish considered dandelion a tonic and a remedy for diseases of the heart. Perhaps a bit maudlin from too much dandelion wine, the Irish, reports Lesley Gordon, called dandelion heart-fever-grass because of its reputation as, "a sovereign remedy against swooning and passions of the heart."

It was dandelion's fine reputation as a physic that earned it passage across the Atlantic to America. Puritan settlers in New England carried dandelion and other seeds with them for planting in family herb gardens on the edge of the wilderness. French, English, Dutch and German settlers valued it for food and especially as a spring tonic after bitter northeast winters. German settlers in Pennsylvania celebrated the arrival of spring with a ritual "thinning the blood," which included dandelions in a tonic. They also mixed dandelion juice with whiskey for *schwitzgegrieder* or "sweat herb."

The noted American herbalist Joseph Smith valued dandelion for its laxative qualities and its capacity to remove obstructions by opening the passages and pores of the body. Taken as a tea or syrup, Smith advised, it "opens the system in general." Native herbalists, widely knowledgeable in the medicinal use of plants, were quick to seize upon this useful newcomer. They steeped the leaves for a physic, and prescribed dandelion tea for heartburn, chest pains and stomach upset. When the first issue of the U.S. *National Formulary* appeared in 1888 with a listing of 435 medicinal formulas, a full dozen of them included "a compound elixir of *Taraxacum*."

* * * * *

Once it had achieved a toehold in the New World amid the stern righteousness of Puritan New England, dandelion was quick to make its escape. It leapt from herb garden to lane and from lane to meadow. Then it was away, wild and carefree once again, appearing in woodland clearings, popping up unaccountably in the wildest places. By the time of the Civil War, the plant was sufficiently widespread that embattled Confederate armies were using its dried roots for coffee substitute, and both sides were using it as a medicinal herb.

Along with other free spirits, dandelion hopped across the vast midwest and prairies, mingling happily with native wildflowers and grasses. Wherever settlers went, plodding along in ox carts or horse-drawn wagons, dandelion went with them, sailing freely on the breeze, sometimes greeting them upon arrival at the homestead, sometimes blowing in a season or two later like the family black sheep, keeping its own eccentric schedule and wandering wherever it would. Earthbound homesteaders accepted it happily as a boon of free food, medicine and wine. Canadian pioneer journalist Susanna Moodie recorded in her journal of 1852, "the dandelion, planted in trenches and blanched to a beautiful cream colour with straw makes excellent salad, quite equal to endive, and is more hardy and requires less care." The Apache and other western aboriginals are said to have welcomed its arrival and given it pride of place at their spring rituals. By the dawn of the nineteenth century, dandelion was well established on the Pacific coast, having crossed the continent and completed its global tour of earth's temperate regions.

And then the trouble began.

A weed, my Oxford dictionary tells me, is a herbaceous plant not valued for use or beauty, growing wild and rank, and regarded as cumbering the ground or hindering the growth of superior vegetation. I imagine that if you were to magically silence all the lawn mowers on a Saturday afternoon in suburbia, and ask the startled home-owners to name the most noxious of weeds, many would name dandelion. How, you might ask, can this delicacy of goddesses have sunk so low in public esteem that pesticide manufacturers can bombard us with springtime television commercials showing dandelions wilting and dying amid otherwise perfect greenswards?

Here's a plant which has fed and healed humans for thousands of years. Here's a bloom of such exquisite beauty that the Japanese formed a national Dandelion Society and hybridized more than two hundred varieties in striking blooms of pure white and deepest black, burnished copper and orange. Now we see it poisoned and despised because it dares defile the manicured perfection of America's Savannah Syndrome. Dandelion became a world traveller through its exuberance in propagation, and that enthusiasm has brought it head to head with a similar urge in humans, to propagate excessively and to control the ground we claim.

But dandelion has claims of its own and is superbly adapted to advance them. Its golden flower head is not one flower, but many clustered together. Each "petal" is actually a flower unto itself, composed of five fused petals. So when the plant sets seeds, it does so in abundance. The puffball that we played with as kids might have held fifty or more seeds, each with its own little parachute. Dispersed by the wind, the tiny parachutes often travel great distances before alighting. Sustained by a thick, fleshy taproot which draws moisture and nourishment from deep in the earth, the mother plant continues producing flowers and seeds throughout the season, scattering hundreds of parachuting pioneers into the warm breezes of summer.

If a single wind-blown seed should land and germinate in some remote spot, far from other dandelions, the new plant will in turn produce seeds, even though it has not been pollinated by a neighbour. For dandelions, like certain other weeds, have learned to get along without sex completely. They have become parthenogenetic—from the Greek for "virgin"—that is, able to develop seeds without fertilization. The tiny seeds that go parachuting off into the breeze will, if they survive, produce perfect clones of their mother.

"Dandelions are often exclusively parthenogenetic," writes Adrian Forsyth in *A Natural History of Sex*. Forsyth describes how asexual dandelions were brought to Greenland from Iceland with Viking colonists a thousand years ago. After five hundred years of bare survival in harsh conditions, he writes, "the isolated colonists had slowly died out. The Greenland dandelions, however, persist today, recognizable in every botanical character as identical to the Iceland dandelions." Forsyth says that the ecological feature which characterizes most parthenogenetic plants is weediness: "Asexuality is favoured where growth and colonization ability are at a premium."

* * * * *

Where I live, dandelions don't seem to be a first-come pioneer species. Freshly disturbed earth will be quickly colonized by thistles, nettles, chickweed, fireweed and other opportunists, whereas dandelions seem to prefer a more settled situation. They happily root in narrow cracks in our flagstone pathways, where they're all but impossible to uproot. They will, warns the British Crop Protection Council, "grow on stone and ash paths and produce seed to infest the rest of the garden." And lawns. Dandelions, as every suburban grass-grower knows, love to settle into lawns, especially where grass is sparse and there's enough space for a seed to shoulder its way in. "Let one into your lawn," warns one observer, "and you are in trouble."

A small colony of dandelions can be controlled mechanically by cutting off the tops before flowers form and digging out the taproot. But you must get every bit of the root. "In disturbed ground," advises the British Crop Protection Council, dandelions "can regenerate when buried or even when left upon the surface. If the taproots are cut up into fragments . . . dandelions can produce shoots from all portions." I've read of one compulsive gardener who dashed about his demesne with a vacuum cleaner, sucking up blowballs before the seeds had an opportunity to disperse.

Lesley Gordon quotes the American garden writer Alice Morse Earle who described, in 1911, an early and effective method of controlling unwanted dandelions in the parks of eastern cities.

It is always interesting to see, in May, on the closely guarded lawns and field expanses of our city parks, the hundreds of bareheaded gayly dressed Italian and Portuguese women and children eagerly gathering the young Dandelion plants to add to their meagre fare as a greatly loved delicacy. They collect these greens in highly colored handkerchiefs, in baskets, in squares of sheeting; I have seen the women bearing off a half-bushel of plants; even their stumpy little children are impressed to increase the welcome harvest, and with a broken knife dig eagerly in the greensward. The thrifty park commissioners, in dandelion-time, relax their rigid rules, KEEP OFF THE GRASS, and turn the salad-loving Italians loose to improve the public lawns by freeing them from weeds.

Nowadays thrifty park commissioners may prefer a quick dose of herbicide to get rid of unwanted weeds. The B.C. *Handbook for Pesticide Applicators and Pesticide Dispensers* recommends the herbicide 2,4-D for control of dandelions. Sold under a variety of trade names such as Weedex, Weed-B-Gon, and Formula 40, this phenoxy acid compound is what's called a "selective herbicide"—it will quickly kill broad-leafed plants but not harm grasses. This makes the stuff seem ideal for lawns, cereal crops and pastures. In fact, it's the herbicide of choice across the continent and millions of litres of it are sprayed on North America every year.

One of the ingredients of "Agent Orange" Vietnam infamy, 2,4-D has a very dark side, and there's a compelling body of evidence that it is a dangerous substance whose continued use is due solely to the political clout of manufacturer and user groups. Several years ago I interviewed John Warnock, author of *The Other Face of 2,4-D,* who is convinced that the herbicide represents a grave threat to public and environmental health. Referring to the fact that 2,4-D was originally registered for use when pesticide approval standards were extremely loose, Warnock told me, "If 2,4-D had to be registered today, I believe it would not pass the standard eight tests for mutagenesis (mutation and genetic changes). Almost all tests, including those done by Dow Chemical Company, show 2,4-D to be teratogenic (causing birth defects). As for carcinogenicity, studies in Sweden conclude that phenoxy herbicides, including 2,4-D, cause cancer in humans."

Despite such evidence, and despite various "extensive reviews" by government and corporate scientists, 2,4-D remains readily available to any homeowner who wants to get dandelions out of the lawn. It would, I was told by a representative of Health and Welfare Canada, be "irresponsible" to remove 2,4-D's registration just because it doesn't meet today's more stringent standards for approving new pesticides.

But how "responsible" is it to splash this poison around indiscriminately, particularly using it for such non-essential uses as killing dandelions in lawns? A California research team outlines a different approach in a report on school ground weed management prepared for an Oregon school district. First, the report suggests, "re-examine community and staff tolerance levels for 'weeds' in turf areas." Even though conventional maintenance standards view all broad-leaf species in turf as undesirable, the report points out, "many common lawn weeds such as dandelions and English daisies are considered quite attractive by many people and play a useful role in improving both soil fertility and aeration."

The report goes on to suggest identifying weed tolerance levels for different turf areas, such as lawns, playgrounds and playing fields. In areas where weeds are a problem, it recommends a program of sophisticated turf maintenance, including frequent aeration, top-dressing with composted sludge or other organic material, overseeding weakened turf areas with drought-tolerant and self-fertilizing grasses, improved watering and cutting schedules, and similar measures. Other mechanical and cultural programs are recommended to control dandelions and other weeds in shrub beds, pavement cracks and around fences and buildings.

* * * * *

"The dandelion," concludes E.R. Spencer in *All About Weeds*, "is a weed, but a virtuous one." To its time-honoured virtues, modern science has added several more. Latex from the plant can and has been used to produce rubber. Organic gardeners have long maintained that dandelions possess the power to stimulate fruits and flowers to ripen, and modern research confirms this folk wisdom. In her book *Green Immigrants*, American writer Claire Shaver Haughton describes how researchers have demonstrated that dandelions exude ethylene gas at sunset, and that mass plantings of them in orchards cause fruit to ripen days earlier and to produce bigger and better fruit.

It has been estimated that dandelions play a part in the ecology of ninety-three different species of insects. Certainly the flowers produce abundant amounts of nectar and pollen, utilized by honey bees for producing golden honey and bee bread to be fed to larvae in the hive. Claire Shaver Haughton says that were it not for dandelion, many wild bees would fade into extinction because our sloppy urban sprawl has obliterated vast fields of wildflowers upon which they once fed. Now dandelion blooms just as profusely among the cracks and crevices of civilization, and wild bees thrive upon it.

More whimsically, dandelion seems closest to us when we're young or in love. Who among us hasn't told the time by the number of puffs it takes to blow all the seed-heads from a stalk? If a child can blow all the seeds away in three puffs, says another bit of childlore, the child's not wanted by its mother; if some seeds still remain, it's best to run home. Lovers can tell how much their beloved is thinking of them by the amount of down remaining on a blowball after a single puff. Or they can transmit messages of love by blowing the down in the direction of their beloved.

In a world of grey flannel suits, snarled traffic and ubiquitous concrete, the tramp with the golden head remains a capricious reminder of what it's like to be young, to be in love, and—oh, glorious!—to be free!

Chapter Seven

PIGEONS

Rock Doves on a Roll

W oody Allen once called them "rats with wings." An Audubon Society official described them as "an ecological nightmare." And a former president of the American Pigeon Racing Union grumbled, "They're mongrels. We're trying to eliminate them. They give us a bad reputation." All three were talking about pigeons—the common domestic variety that flocks in city parks and squares. A dirty nuisance to some people, beloved friend to others, to some a bird synonymous with stupidity, to others a creature of surprising intelligence—the pigeon struts and fluffs and coos its way through human consciousness, arousing all sorts of fierce passions. Even as mild-mannered a fellow as Jesus Christ, entering the temple at Jerusalem, "over-turned the tables of the money-changers and the seats of those who sold pigeons."

Why all this passion? Pigeons are possibly the most familiar bird on earth, with their brilliantly iridescent purple and green feathers, their wing-clapping flight and stuttering coos, their eyes of fiery orange. For many inner-city dwellers they're a lifeline to a natural world all but eradicated by highrise buildings and traffic-choked streets. They serve as companions to the poor and dispossessed and as a steady metaphor for down-and-out urban poets.

Still, they carry a huge burden of contempt. The name "pigeon" does them little good, with all its connotations of deceit and deformity—stool pigeon, pigeon-toed, pigeon-chested and all the rest. More properly they're called the rock dove, which has a lovelier ring to it and far sweeter connotations. A cliff-dwelling bird native to the Mediterranean and Middle East, its formal name is *Columba livia*. Wild populations of rock doves still live on sea cliffs of the Old World, but today they're vastly outnumbered by millions of pigeons, descended from the ancient wild stock, and bred by humans into hundreds of specialized varieties. Common street pigeons have simply reversed the process, being feral descendants of domesticated birds which were in turn bred from the wild doves. These feral mongrels now colonize the crannies and crevices of urban buildings the way their ancestors did high cliffs.

"Many persons have quite a mania for pigeons," wrote the Roman Pliny two thousand years ago, "building houses for them on the tops of their roofs, and taking delight in relating the pedigree and noble origin of each." Domestication of the rock dove is thought to have started at least five thousand years ago in the Near East. Egyptians of the Old Kingdom were raising pigeons for food three thousand years before Pliny's time, and the domestic breeds known today had their origins in the yard birds of the Near East. Much later, the groaning banquet tables of mediaeval English nobles included "squabs"—young pigeons about to fledge—and these have continued to be a highly prized delicacy. By the 1920s huge pigeon plants were established in the eastern U.S., some housing thirty thousand breeding adults, to supply the luxury trade in squabs. In the mid-eighties, French and Italian gourmands were consuming some 90 million squab a year, mostly prepared by baking or broiling. In Hong Kong, where deep frying is the preferred preparation, more than 5 million birds are eaten annually. Comparable to chicken, the meat is described as more tender and more succulent.

* * * * *

But, like Pliny's roof-top show-offs, many people have kept and bred pigeons for nobler reasons than stuffing one down for supper as though it were Kentucky fried chicken. Of the more than three hundred subspecies known today, many are bred and pampered as show pigeons, homing or racing pigeons. And here one glimpses a refined world of wealthy aviculturalists, willing to spend many thousands of dollars on a single breeding pair. At national pigeon shows avid fanciers gather to cluck and coo about elegant Jacobins, Russian trumpeters and aristocratic English carriers. "Racing homers" are pedigreed birds bred for speed and a highly refined homing instinct. Capable of sustained flight at seventy kilometres per hour, they were once the "winged telegraphs" carrying news of victory in battle, of pestilence or disaster in the fastest available method of long-distance communication.

As passionate about their birds as any pigeon-fanciers, southern European immigrants to the big industrial cities of America brought with them the "pigeon game" or "thieving." They'd house their prized pigeon flocks on apartment roofs much as in Pliny's day. Owners on adjacent roofs would then release their flocks simultaneously, so that the birds would flutter into the sky and intermingle. The game's object was to entice the neighbour's birds to return to your cages where they could be held for ransom. The unwritten rules of the sport—predominantly a male preoccupation—permitted successful handlers to heap scorn and derision upon the owners of birds they'd "thieved."

Racing homers have the best-developed homing instinct of any pigeon, but the instinct runs strong even in park mongrels. Municipal authorities in Paris failed to account for this when, with great to-do, they netted thousands of pigeons in the centre of the city and trucked them off two hundred kilometres away. Released in the countryside, the Parisian pigeons promptly took flight and were back to the capital before their red-faced captors!

Scientists are still scratching their heads about how pigeons manage to return so unerringly home. There's an ongoing debate about whether they employ earth's magnetic fields in some way. One theory has them using the moon for guidance. Other explanations of the bird's legendary navigational skill credit a combination of acute vision and superlative memory for topographical detail.

"Pigeons are not just opportunistic creatures like rats," University of Iowa psychologist Dr. Edward A. Wasserman told the *New York Times* a few years ago, "they're really part of the human environment and they have some striking features in common with us, acute vision for one." Those bright and beady pigeon eyes are actually much better than ours, producing a stereoscopic image of objects straight ahead, as well as a wide-angle monoscopic view of the periphery. Just how good they are was demonstrated by a recent U.S. Coast Guard plan named "Project Sea Hunt." The idea was to exploit the pigeon's acute sight and its learning ability in order to spot shipwreck survivors at sea. A three-pigeon squad of ordinary street pigeons was trained to peck a key upon spotting an orange life jacket. In eighty-nine trial flights the pigeons, suspended in a transparent box beneath a helicopter, spotted an orange buoy 96 per cent of the time, sometimes from as far away as seven hundred metres. The human flight crew, by comparison, had a meagre 35 per cent success rate. Though promising, the project was eventually scrapped due to budget cutbacks.

During World War II, when carrier pigeons were still being used as messengers, trained pigeons were seriously considered as cheap and replaceable guidance systems for antisubmarine missiles. A three-pigeon crew, encased in the nose of a missile behind three little windows, was expected to guide the missile, kamikaze-style, into its target. Perhaps recalling the pigeon's age-old reputation for cowardice—melancholy Hamlet at one point moans, "But I am pigeon-livered and lack gall"—the Pentagon cancelled this program too.

As with a certain type of military mind, the pigeon has an established reputation for stupidity, based upon the ease with which it can be snared. "A cat among the pigeons" denotes a shrewd operator loose among the rubes. "To pluck a pigeon" is to cheat a gullible person of their money. And in sporting circles, to "pigeon" someone was slang for cheating by a fairly transparent hoax.

But Dr. Wasserman disagrees, arguing that "the conceptual abilities of pigeons are more advanced than hitherto suspected." With a brain smaller than a fingertip, a pigeon can perform certain tasks still beyond the capacity of any computer. "Pigeons commit new images to memory at lightning speed," he says, but the remarkable thing is that they organize images of things into the same logical categories that human beings use when we conceptualize";

by studying how their minds work, "we are getting closer to knowing what intelligence is and how it came into existence."

* * * * *

Considering how humans have dealt with pigeons over the years, one wonders whether "intelligence" isn't too strong a word. Hallucinogenic drugs, plastic snakes and fistfights have all contributed to the war against these mongrels. In B.C., as elsewhere, they're listed as a nuisance and crop-damaging bird which may be killed without a permit. The pigeon is, in the words of writer Paul Galloway, "a prototypical urban dweller—it's noisy, pushy and a polluter." Most problems arise from their droppings, the irreverent bane of park statuary everywhere. The droppings are highly acidic, capable of disintegrating cement and eating through paint on buildings and automobiles. A few years ago municipal authorities in Venice—home to some 300,000 pigeons—and Florence began trapping pigeons because the cities' magnificent masonry and statuary were being eaten away by pigeon dung.

"Bird droppings contaminate food and water," warns my pesticide handbook. "Ectoparasites such as fleas, ticks and mites invade buildings from roosting sites. Rats and flies are attracted to pigeon roosting areas and may be further disease vectors." The list of pigeon-borne diseases is long and scary enough to keep you out of city parks forever, and helps municipal health authorities to whip up "pigeon fever" at whim. Writes W.F. Hollander: "A succession of such scares has emanated from the great cities. First there was the salmonella scare of the 1930s. Then equine encephalitis, psittacosis-ornithosis, Newcastle disease, toxoplasmosis, and finally cryptococcosis." When sixty-eight residents of Windsor, Ontario, contracted encephalitis in 1974, and five of them eventually died from the disease, blame was laid on mosquitoes carrying the infection from pigeons.

Responding to the threat, authorities have tried to control pigeons with attempts ranging from the cruel to the comic. Weather-protected ledges on buildings make perfect pigeon roosts, and discouraging the birds has led to experiments with twisted chicken wire and intense fluorescent lighting. In Oklahoma City, four dozen vicious-looking plastic snakes were placed along the limestone ledges of the state capital building in a whimsical attempt to scare off roosting pigeons. Others have tried using obnoxious pastes, with brand names like Shoo-Bird and Tacky-Toes, which come in tubes like caulking. A bead of the stuff laid along the ledge will burn the feet of any bird alighting there. Renewed every week or so for about a month, these have proven effective in at least temporarily forcing the birds to roost elsewhere.

A more sinister variation on the glue-paste theme is an avicide named Queletox, a highly toxic substance which, absorbed through the bird's feet, kills it within a few hours. Approved for use against pigeons and starlings, this lovely stuff comes with the warning: "May be fatal if inhaled, swallowed or absorbed through the skin." Other chemicals have been tried which, mixed with food, aim to disorient the birds or inhibit their breeding.

In 1986 the Department of Public Works in Ottawa tried to scare pigeons away from the National Conference Centre and other public buildings by feeding them the chemical Avitol. After eating it the birds were reported to "squawk, screech and swoop about in a drugged frenzy," which the Department hoped would give fair warning to other pigeons. At least one city in the southwest has tried putting out food laced with anaesthetic and then feeding the comatose birds to zoo animals. Rather then attack the pigeons, a cautious Vancouver city council voted to issue tickets to pigeon-feeders under an obscure anti-littering bylaw. New York City dabbled with a contraceptive program, and has come up with the most elegant, if only partial, solution—a dozen pairs of peregrine falcon which nest high on the clifflike office towers and swoop down in awesome power dives to snatch helpless pigeons for food. San Francisco tried a plastic substitute, placing fake owls to terrify their pigeons. Eventually, and predictably, the pigeons took to perching on the owls and streaking them with droppings.

Municipal politicians in pigeon-plagued cities have learned to step lightly through the pigeon problem for fear of provoking the wrath of the "pigeon lobby"—a loose coalition of bird lovers, park goers, deep ecologists and animal rights activists which any city council arouses at its peril. Take the case of Chicago, which established a program of trapping and gassing ten thousand city pigeons a year. Even the local Audubon Society split on the issue. "The pigeon does no good and it does some harm," said the chapter vice-president, "but many people and a fair number of our members are sentimental when it comes to wiping out one segment of our fauna, even if it is alien." The *Chicago Sun-Times* reported that professional exterminators were refusing to even talk about the pigeon kill. "You can't blame us for not wanting to talk," one pigeon exterminator told reporters. "Whenever an article appears in the paper, we'll get calls at two or three in the morning. 'You pigeon killers!' they'll scream, 'You murderers!' "

Things got really out of hand in one small Pennsylvania town in 1990 at an annual pigeon shoot where caged birds are released and then shot by paying customers. Fourteen protestors with fake blood dripping from their mouths raced onto the field and tried to free the pigeons. Fistfights broke out, a windshield was kicked in, a dead skunk was hurled at the demonstrators, three state troopers were injured and twenty-five people arrested, with the crowd chanting "Shoot! Shoot!" as police tried to handcuff the protestors.

* * * * *

All this sound and fury seems a long way from the gregarious and gentle birds that strut and coo and scramble for bread scraps in our city parks. Donald Stokes, author of *A Guide to Bird Behaviour*, recommends in his first volume close observation of *Columba livia* as a good introduction to understanding how birds communicate through visual and auditory displays. He identifies eight specific visual displays that pigeons use, three of them accompanied by particular calls. Most of these activities are related to pair-bonding, mating and nest-building, and Stokes says pigeons engage in at least six conspicuous courtship displays through most of the year.

The preliminaries of courtship take place on the feeding grounds. Males entering or leaving the flock, writes Stokes, may engage in wing-flapping while in flight, possibly to advertise their sexual maturity. The wing-flapping "stimulates males and females to also take flight with wing-flapping." A male in search of a mate may land near a feeding group and engage in "bowing," in which he ruffles his neck feathers, lowers his head and rotates in half-circles. Between bows he may try "tail dragging," in which he lifts his head high and does a short dash with his spread tail feathers dragging on the ground, while he utters a distinctive little chorus of quiet coos.

After a pair is bonded they may engage in "driving," both of them running through the flock with the male appearing to drive the female along. In the last week before egg-laying, the pair switch to nesting and mating displays, including "nodding," in which a bird crouches low and repeatedly nods its head as though pecking, while giving a long and drawn-out coo. Just before mating the pair may try "billing," in which the female puts her bill into her mate's open mouth and the two of them move their heads rhythmically up and down.

Nest-building and incubation involve both parents. It used to be thought that pigeons always hatched one male and one female chick, and the old expression "pigeon pair" was applied to boy and girl twins. The clutch hatches in eighteen days, and for the next ten days both parents feed the nestlings, first with food regurgitated from the crop, called "pigeon milk," gradually switching to fruit and insects. Survival rate for city chicks is about 70 per cent, and pigeons live up to fifteen years. Mates for life, they may raise two or three broods between February and August. By late August or early September the adults undergo a complete moult. Flocks have no real seasonal movement patterns, and pretty much stick to the same general area throughout the year.

* * * * *

We never see *Columba livia* around our place—maybe the hawks and owls in our woodlands are enough to keep them away, or maybe they're just too citified for our kind of rough country. Instead, we get band-tailed pigeons, *Columba fasciata*, in the yard, especially in summer when pairs of them flap in to feed on elderberry fruits. The locals call them "wood pigeons." Large and beautiful blue grey birds, they like to perch high up on tree branches before swooping down for a feast of berries. When they take off out of the bushes, there's always a great clatter and flapping of wings. They're common all down the west coast from southern B.C. to Mexico, and in places are shot as a game bird. At one time excessive killing of band-tails was thought to be threatening species survival, but happily their numbers have rebounded in recent years.

North Americans, of course, are heirs to one of the truly colossal monuments to interspecies stupidity—extinction of the passenger pigeon. Before the arrival of Europeans, these magnificent creatures numbered in the billions. Ornithologists speculate that these swift fliers, believed capable of speeds up to 112 kilometres per hour, composed between 25 and 40 per cent of the total bird population of the continent! Possibly the most numerous species of bird that ever existed on earth, the sky literally darkened for hours as flocks of them passed overhead.

Slaughtered indiscriminately for "sport" and for the popular pigeon pie, within a century these tremendous flocks were gone. When the very last of their number died in a Cincinnati zoo in 1914, we bid farewell forever to what has been called, "the most impressive species of bird that man has ever known."

In its stead we now have millions of mongrel pigeons crowded, like ourselves, into cities, scratching out a living, streaking our monuments with excrement. Sometimes that seems precisely what we deserve.

Chapter Eight

FLIES
Awful Fecundity

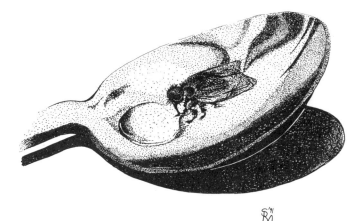

When a worker from the local salmon farm called me one winter morning to ask if I'd like a truckload of dead fish ("morts" in fish farmer lingo), I replied yes, of course. Hungry seals had been breaking into the pens and killing large numbers of captive salmon. Retrieved by divers, the salmon corpses can pose a real disposal problem for fish farmers—unless there are avid gardeners like ourselves who'll eagerly accept the fish for use as fertilizer.

The seal predations grew worse that winter, and in the end some fifteen thousand salmon—and a shocking number of seals—were killed at the farm. Truckloads of dead fish kept arriving at our composting area. At the same time I was donated several truckloads of fresh horse manure, in the field. With tremendous gusto I set about constructing compost heaps. I'd spread a layer of manure, then a layer of fish, sprinkle the fish with hydrated lime, add a layer of spoiled hay, and repeat the sequence until the heaps were almost two metres high. Then I covered them with black plastic and left them to work, confident that by summertime we'd be lavishing fabulously rich compost on vegetables and ornamentals.

Within days, the piles began to heat up, sending off shimmering gases and that particular aroma so distasteful to the prim, but to the gardener so redolent with grand possibilities. During the first warm spell of March, I began to notice flies. A few dozen of them would gather on top of the black plastic, basking in the pallid sunshine. I paid them no attention, other than to think it was a strange time of year to be seeing flies. As the heaps cooled down from want of oxygen, and the reluctant West Coast spring tiptoed in, the flies began to multiply. I knew it was time to turn the heaps and get them heated up again with fresh oxygen, but other chores somehow pressed more urgently, and I put the job off. Then I really began noticing flies. On warm afternoons there'd be scores of them gathered on each pile.

It was a mild spring day in April when compost-turning finally struggled to the top of my chore list. I pulled on my coveralls and gumboots, got my manure fork from the shed, and headed briskly to the first of our four heaps. At my approach a swarm of disgruntled flies buzzed up into the air. I rolled off the rocks pinning down the plastic covering and peeled the sheet off the pile. And there I saw one of the most disgusting sights imaginable—the entire heap, a metre high and two metres in diameter, was a swarming mass of maggots! Not a sprinkling of maggots, not a scattering of maggots, but maggots by the thousands, huge and white and wriggling, several centimetres deep!

As I watched in horrid fascination, the entire mound of them seemed to boil in certain spots, like porridge in a pot, all wriggling frantically, and quickly disappeared inside the compost pile. Within moments they were gone completely. Had this been some bizarre hallucination, some disconcerting side-effect of too much rhubarb wine? No, I could still hear the hordes inside the heap, producing an awful gurgling, mushy sound. A shudder of disgust shook me.

I checked each of the other piles and the same repulsive scene repeated itself. After the maggots had disappeared into the last pile, I plunged my fork into it, and turned a forkful over. The forkload seemed more maggots than compost. Mingled with the maggots were hundreds of small, hard empty shells, and equal numbers of stunned-looking flies, damp and disorganized, wandering over the heaps.

I ran to my shed and grabbed a full bag of hydrated lime. "This'll fix 'em!" I said to myself with the bravado of the true countryman. I spent the next two afternoons turning the soggy heaps and sprinkling each exposed forkful of maggots with lime. They wriggled frantically at its touch, in what I took to be death throes. "Don't worry," advised my organic orchardist friend Larry, "once the piles heat up again, the heat will kill them all."

* * * * *

Dubious, and decidedly paranoid about the several million maggots apparently maturing in the heaps, I set about educating myself on the topic of flies. The best source I could lay my hands on was a book titled *The Natural History of Flies* by British entomologist Harold Oldroyd.

Breeding in dung, carrion, sewage and even living flesh, flies are a subject of disgust, writes Oldroyd, "in fact, flies are a topic like drains, not to be discussed in polite society."

There are some eighty thousand species of them worldwide—and that doesn't include other insects, such as dragonflies, mayflies and fireflies, which aren't true flies at all. But there are humpbacked flies, stiletto flies and flat-footed flies, all of them distinguished from other insects by having only a single pair of wings—the front pair of the double set most insects have. In place of the back pair, a true fly has two small knobs, called halteres, which act like little gyroscopes to keep the fly balanced. Remove them, and the creature falls to the ground incapable of flight.

These distinctions were not foremost in my mind as I pored through Oldroyd's pages. I wanted to know what these multiplying monsters in our compost heaps were and what I should do about them. But first I needed to know that there are three great groupings of flies, based on their evolutionary levels: the primitive ancestral flies like crane flies, mosquitoes and blackflies; an intermediate group including types like robber flies, bee flies and horseflies; and a third group called Cyclorrhapha, which includes the common housefly and the bluebottle. Oldroyd describes these as "aggressively advanced" types, "compact, bristly, with broad wings buzzing at high speed, flight persistent and purposeful." Now, I sensed, I was nearing the answers I sought.

Most of the flies we encounter belong to a group called the Muscoidea, which includes the common housefly (*Musca domestica*), the best-known and perhaps least-loved of all the flies. A master of adaptation, it flourishes wherever humans are to be found, and so is called a "synanthropic" fly. Closely related is the lesser housefly (*Fannia canicularis*). These are the little nuisances which circle endlessly playing "kiss-in-the-ring" in a room or in the shade of a tree. The circling flight, says Oldroyd, is one of the variations of swarming/nuptial flight behaviour.

Another pesty muscid species is the maddening "face fly" (*M. autumnalis*) which looks like a housefly. A native of Europe and Asia, it was introduced to North America only in the 1950s but is now a common pest swarming around the faces of livestock. The equally infuriating "sweat flies" which constantly alight on your body when you're digging or doing other sweaty outdoor things are also muscids; they're of the genus *Hydrotaea*, and as the name implies, they like to slurp up perspiration from your skin.

None of these flies bite or sting. Instead, they feed by sponging up liquids through complex mouth parts. If the foodstuff isn't liquid to begin with, they may liquefy it by regurgitating upon it, the unlovely phenomenon of "vomit drop." They're drawn to dung of all sorts, including human excrement, and lay their eggs in it or in rotting vegetable matter, upon which their maggots can feed.

Now I was getting warmer in my search; but so was the weather and so were our compost heaps. The anticipated kill-off from the heating heaps had not occurred; quite the reverse, in fact, and throughout our yard swarms of new flies were gathering menacingly. Visitors were beginning to pass comments.

* * * * *

I pressed forward with my researches, and was soon rewarded. The maggots in my heaps, and by now I had determined there was more than one kind of maggot present, were those of the so-called "flesh flies," the carrion-feeders whose larvae live primarily on decaying flesh. Entomologists classify them in a number of ways, but I quickly zeroed in upon the two families to which I suspected my unwanted colonies belonged: the blow flies (Calliphoridae) and the flesh flies (Carcophagidae). Common blow flies are coloured metallic blue or green and resemble houseflies. The most familiar of them are the bluebottles—one type of which has the graphically descriptive formal name *C. vomitaria*—and the greenbottle (*Lucilia sericata*). Flesh flies look even more like houseflies and lack the bright colours of the blow flies.

Many of these species will only deposit their eggs in recently dead flesh, and their frantic buzzing-about is often a desperate search for a suitable corpse upon which to deposit their bulging eggs. In some species the eggs hatch within the female's body and the maggots are deposited directly onto dead flesh. At one time it was believed that blow flies literally "blew" their maggots onto carcasses. "In suitable weather," writes Oldroyd, "arrival of the egg-laying females is prompt. Hatching of eggs and survival of larvae then depend upon sufficient warmth and humidity." My carefully constructed compost heaps, I realized, were ideal breeding grounds, veritable incubation chambers, providing warmth from the composting process, abundant dead flesh for food and ample moisture from the rotting fish and dung.

As repulsive as it might seem to us, and so evocative of our own death and decay, the maggot is actually quite an advanced type of larva. With its pointed head and blunt rear end, it wriggles about lacerating rotting flesh by means of a mouth-hook. In his *Studies on the Nutrition of Blow Fly Larvae*, R.P. Hobson describes how the maggot's excrement contains high concentrations of ammonia and protein-digesting enzymes which work to dissolve animal muscle fibre and intramuscular tissues, making them available as maggot food. When liquefaction reaches a certain point, Hobson says, the corpse ceases to be suitable for blow fly larvae and is colonized by the larvae of other flesh flies.

After gorging themselves for a couple of weeks, the maggots abandon their food source and search out a dry, sheltered place in which to pupate—their last instar, or quantum growth leap. In my compost heaps they simply clustered in conveniently dry and warm pockets of hay within the heap. Rather than spinning the silken cocoon of many insects, the maggot simply retains its larval skin, which hardens and darkens into a bullet-shaped outer shell. This little cocoon—called a puparium—is what I was seeing by the hundreds in my heaps. "When metamorphosis is complete and the young fly is ready to break out," writes Philip Street in *Animal Reproduction*, "it has to rupture the puparium as well as the normal pupal skin." To do so it pumps pressurized blood into a special frontal sac at the tip of its head. Pressure from the expanding sac pops the top off the puparium and allows the adult fly to emerge. The little sac has served its only role in the fly's life and is withdrawn. Meanwhile, the fly, damp and bewildered and unable yet to fly, wanders about the breeding grounds.

* * * * *

By this time our yard was beginning to resemble the land of Egypt during the fourth great plague before the Jewish exodus, when "there came a grievous swarm of flies into the house of Pharaoh, and into the servants' houses, and into all the lands of Egypt; the land was corrupted by reason of the swarm of flies." Given the right conditions, no divine intervention is required for flies to achieve these numbers, for they're capable of awesome multiplication. In temperate regions it only takes three weeks for a fly to develop from egg to breeding adult, and in tropical heat the pace may be accelerated to a single week.

One zoologist has calculated that if all the eggs laid by a common housefly hatched and the young survived through six generations, the offspring, pressed together at 200,000 flies per 0.28 cubic metres of space, would occupy 70,000 cubic metres! Another number cruncher calculated that the progeny of a single pair of houseflies, over the course of one summer, if all hatched and survived, would be 191 followed by eighteen zeroes. Commenting on the proverbial fecundity of houseflies, Oldroyd repeats a published claim that the offspring of one pair in one summer could cover the earth to a depth of just over fourteen metres! "Incredulous," says Oldroyd, "I recalculated this, and decided that a layer of such a thickness would only cover an area the size of Germany: but that is still a lot of flies."

In reality, many of the larvae die from poor weather or are eaten by birds and other predators. I took some heart from this. But the bad news is that most flies spend their whole lives within about 1.5 kilometres of their breeding place. Dispersal was a vain hope for me. Furthermore, the threat that flies pose to human health is awesome: the common housefly is known to transmit some thirty different diseases and parasites to humans, mostly by feeding on infected material and subsequently alighting on human food and soiling it. The list reads like a rogue's gallery of medical miscreants: leprosy, dysentery, diphtheria, smallpox, typhoid, cholera, scarlet fever and many more. Notwithstanding our friend the flea, some observers regard the common housefly as potentially the most dangerous insect in the world.

And although blow flies and flesh flies are less of a menace to humans, because they don't as readily come indoors and don't feed on as wide a variety of human foods as houseflies, still my exponentially rising population plainly represented a potential health hazard. By now, swarms of flies were blackening fenceposts and covering whole walls of outbuildings. Friends were refusing to visit. I was beginning to panic. I abandoned the ineffectual heat-and-lime treatment and sought more drastic measures.

* * * * *

Perhaps beguiled by those TV commercials in which bewildered young couples find answers to their house and garden dilemmas in the sage advice to be found at the "home of the handyman," I went to a nearby hardware store. I explained my situation to the clerk. Confidently she escorted me to the pest-control section. "We have these," she said, holding up a dainty little roll of flypaper. I almost burst into guffaws—all of the rolls together wouldn't begin to dent my awesome populations. "Anything on a bigger scale?" I asked. "You can always use Raid," she replied with a helpful smile, showing me the familiar aerosol canister. I fled the place.

Back home, I consulted my trusty *Handbook for Pesticide Applicators and Pesticide Dispensers*. It calmed my anxieties with the reassurance that there are many effective types of insect control. Chemical control, for example. When DDT and similar products were first introduced, they achieved breathtaking results in reducing fly populations, even in badly infested tropical localities. But fast-breeding flies quickly developed resistance to the insecticides and, as Oldroyd puts it, "houseflies have developed such a high degree of resistance to DDT and related materials that satisfactory control is impossible in most areas." Damn!

Then there was the celebrated case of Larvadex, heavily promoted in the early eighties as a "one-shot, labour-free" fly killer with "no side effects." Just the kind of thing I was looking for. Larvadex was developed as a "feed-through" pesticide and was widely used to combat flies on poultry farms. Advertised as a "whole new concept" that "goes to the source of the problem," the insecticide was simply mixed with chicken feed, and passed through the bird's digestive system, rendering its manure toxic to fly larvae. Hallelujah! But wouldn't you know it, health officials had to go and discover that the treated chickens would give you cancer if you ate their meat or eggs. Embarrassed officials quickly yanked Larvadex off the market. Foiled again!

Another "feed-through" approach has seen the micro-organism *Bacillus thuringiensis* fed in large doses to steers and chickens. Lester Swan reports in *Beneficial Insects* that this pathogen was, "very effective in preventing houseflies from completing their development in the manure. Most of the mortality occurred in the pupal rather than the larval stage of the flies." There have also been encouraging results in cattle feed lots and dairies using the fly parasite *Spalangia endius*. Released in the millions, these tiny parasites lay their eggs in fly pupae and their offspring proceed to consume the developing fly.

Similarly, a couple of specific fungi to which flies are susceptible have shown themselves to be effective control agents. Irradiation has been used to control fly breeding. Chemosterilants and disorienting sex attractants have been tested. Multitudes of dedicated chemists and biotechnicians, I realized, were scrambling to devise the perfect solution to my problem. I took heart, began to nurture a perverse delight in comparative insecticidal "knock-down power"—as though these wonderful substances were prizefighters battling for my entertainment. I went so far as take distant consolation from Beatrice Trum Hunter's noting that the Mexican marigold, hung in dog kennels and cow sheds in what used to be called Rhodesia, proved effective in keeping flies away.

* * * * *

And while rummaging through these exotic remedies, I realized I could as well be cheered that our infestation was really quite benign compared with those nasty foreign species. It's true, Oldroyd reports, that there have been cases of blowfly maggots surviving ingestion by humans. "These strong maggots," he writes, "equipped with powerful mouth-hooks, can set up acute indigestion and nausea, and even cause haemorrhage of the intestinal wall. They do not readily tolerate lack of oxygen, and become violent when they're short of air, provoking retching and vomiting, and so partially emptying the stomach." But such incidents, he reassures us, are rare because the maggots are so conspicuous in "fly-blown" meat they're easily spotted.

How much more intimidating, for example, is the world's largest known fly, a denizen of tropical South America known as *Mydas heros*. With a body length of 60 millimetres and a wingspan of 100 millimetres, this monster, says the *Guinness Book of Animal Facts and Feats*, is "so formidable it will attack even well-armed bees and wasps, diving on their backs and paralyzing them with a bite in the soft region of the neck." Or there's the African tumbu fly, *Cordylobia anthropophaga*, whose larvae wriggle about in a boil-like swelling under the skin of humans. There are malarial mosquitoes and tse-tse flies. And, perhaps most repugnant of all, the Congo floor maggot which creeps about at night sucking the blood of sleeping people.

Compared with such horrors, my own situation seemed trifling indeed, and its eventual solution was a matter of utmost simplicity. It began with an observation from my sister-in-law Barbara, one of several people who had maintained a morbid fascination with our maggots. She expressed to me a perfectly reasonable concern: that by allowing maggots to eat up the compost materials and eventually buzz off as flies, I was actually depleting my supply of compost. But rather than enhancing my panic, this observation gave me a vital clue to solution.

I'd noticed that in cool or wet weather, all the adult flies went back under the plastic covers to shelter in the warmth of the compost. Entrapment became my strategy. I waited for a rainy day, and with all the flies safely tucked in, I quickly sealed their escape routes and smothered the heaps with wheelbarrow-loads of cedar shavings, piled twenty centimetres deep over the plastic. A horrendous buzzing ensued. And then: silence. Within days the few stragglers left outdoors were gone. There was not a fly to be seen anywhere. No sound of squishy wriggling from the maggot-ridden heaps. A blessed contentment descended.

After they were dead, I began to think more kindly of flies. Plainly, their vast numbers feed all sorts of other creatures. Their maggots play an important role in breaking down and recycling carrion. Zoo keepers breed maggots as feed for their caged insectivores, and in some places blow fly maggots (renamed "gentles") are raised as valued bait for anglers. And, best of all I discovered, they make wonderful compost. Yes! In fact, one woman whose magnificent zinnias captured the grand prize at a flower show confessed in the glory of the moment that she'd grown her prize-winning beauties in compost enriched with dead flies. So there I found myself, proud owner of four potentially prize-winning compost heaps, a shrewd fellow indeed, one of whom it could be said proverbially, "there's no flies on him."

Chapter Nine

MOSS

Softening the Stones

O ne of the surest signs of spring in the Pacific Northwest is the annual blossoming of television commercials advertising moss killer for lawns. Garden supply stores bulge with bags of the stuff, dutifully purchased by home-owners determined to eliminate invading mosses from their greenswards. It's a fool's undertaking, of course, a labour of illusion roughly equivalent to establishing one's domicile in the Sahara and setting about eliminating all that nasty sand.

Moss likes it here on the cool, moist western edge of the continent. It's been here for untold thousands of years, surviving in places even through the awesome advances of the glaciers. Moss was undoubtedly instrumental in re-establishing great conifer forests on rock scraped bare by the ice sheets. Bryologists have identified more than twenty-three thousand mosses and liverworts worldwide. About seven hundred of them are known to grow in the region where I live. They cling to boulders, cliffsides and the trunks of trees, tuft like nests on the tops of branches, and spread everywhere across the forest floor. In quaking bogs and swamps they spread, within the splash of waterfalls and by the surge of surf. Their colours range from vivid purple reds to delicate pale yellows and diverse shades of green. They thrive in sunshine and shade, from sea level to high alpine, in the depths of dripping caves and on windswept mountain crags.

Establishing ourselves amid this maelstrom of mosses, we'd be fools to think they'd somehow exempt our holdings from their relentless colonizing of surfaces. And we find no exemption: they alight on our roofs and set to work among the shingles. They clutter up our rain gutters. If there's a bit of exposed stone or concrete, a crack in the wall or a shady crevice, some spore of moss will settle there and send its slender filaments creeping for a toehold. Every summer I spend a couple of afternoons crawling across the roofs of our house and outbuildings, scraping the cedar shakes clean of the year's accumulated feather mosses, certain that by next year they'll be back again, relentless as the rain they love, and as much a part of living here.

"In mosses and lichens," writes John Bland in *Forests of Lilliput*, "strength is mingled with humility, gentleness and charm, with elemental essence, reflecting the gladness of wind, sun and rain." I wish they'd print that on the bags of moss killer, the way health warnings are printed on cigarettes, but I think it'll take a while. Most of us, even those of us living in this profligate garden of mosses, remain oblivious to these strange plants. We might momentarily admire a tumble of mossy boulders alongside a woodland cascade, or curse the mosses invading our lawns, but really we know nothing of what mosses are, where they come from or how they live and multiply around us.

This was precisely my state of mind on mosses—a benign indifference bordering on unconsciousness—until I chanced to meet Dr. Wilf Schofield, a professor at the University of British Columbia specializing in mosses and other bryophytes. There are barely more than a hundred bryologists in all of North America, so I felt myself both lucky and privileged to accompany Dr. Schofield on a moss-hunting expedition ("foray" is the preferred term within the profession) to "the moss capital of the world," the mist-shrouded islands known as Haida Gwaii, or the Queen Charlotte Islands. An internationally recognized scholar, at that time in his late fifties, Wilf Schofield eschews any personal acclaim; he even declined to be photographed for the magazine story I was writing, saying, "The organisms themselves are much more interesting than the people who search for them."

Wilf has been searching for them over the past thirty years, and some of his extensive specimen collections are now housed at U.B.C. Time and again he has returned to the mountain slopes and rain forests of the Queen Charlottes, for that misty archipelago—home of the great Haida culture and breeding grounds for an astounding concentration of seabird and other animal life— is also a treasure house of rare and wonderful mosses, contributing to the islands' being called "the Canadian Galapagos."

On an August afternoon, a battered old float plane deposited us—Wilf, doctoral student John Spence, my partner, Sandy, and myself—on a shingle beach on the west coast of Graham Island. Far beyond the reach of any road, this area has been designated Cameron Range Ecological Reserve because of the rare mosses and other plants which grow here. We set up our camp on the beach and plunged into the forest. Instantly, we were in a magical kingdom, in the grottoes and glades of a dripping rain forest where mosses, liverworts and lichens spread a thickly textured blanket over every surface, whether of earth, boulder or wind-thrown log. This is moss so ancient and luxuriant you can reach your arm deep inside it, as though the earth itself were upholstered with an overstuffed, living softness. Not confined to the forest floor, the plants creep like bright green sheathing up the trunks of spruce, cedar and hemlock trees festooned with filmy lichen banners.

Scrambling about in all this lushness, in bright blue rain gear, hat and flowing cape, poking into crevices with his knife, pulling out scraps of moss, examining them closely under the hand lens dangling from his neck, mumbling indecipherable Latin names, and stuffing samples into small plastic bags, the eminent Dr. Schofield looked like some slightly crazed high priest of bryology. At one point he stopped, peered up at me through spectacles dripping with rain, smiled mischievously and said, "This must look rather silly."

* * * * *

Later, back at camp, we sat around a crackling fire, sipping tea and discussing the day's discoveries. At one point Wilf made the observation that "under the microscope, all things are beautiful." And so it is with mosses and their near relatives the liverworts. Tiny, simple and ancient, these plants create miniature forests of astounding beauty and diversity. In the world of plants they belong to a class of their own, called the Musci, which numbers over twenty thousand species distributed across the earth. Most of them are no more than a few centimetres tall, and many less than a few millimetres.

Mosses are many millions of years old, more primitive by far than many more familiar plants. Scientists who study plant evolution generally consider mosses and liverworts to be among the first plants to have developed on land. It's believed that they evolved from algae, gradually adapting from an aquatic to terrestrial lifestyle. The fossil record indicates that these plants have changed very little since their ancient beginnings. Bland writes, "Some botanists believe mosses and liverworts enjoyed their greatest prosperity during the Devonian and Silurion periods, 376 to 400 million years ago."

A number of the plants we call mosses—reindeer and sea moss, Spanish, Irish and Scottish moss—aren't mosses at all. When Longfellow waxes poetic about "bearded with moss and in garments green," he really means the lichen we call old man's beard, which is not a moss. True mosses differ dramatically from most other plants, since they don't have roots and don't produce seeds. Instead of roots, they're anchored by tiny filaments called rhizoids. Nor do they feed like other plants, which draw nutrients up from the soil through a vascular system connecting roots, stems and leaves. Instead, mosses absorb nutrients directly into their leaves from water, soil and air, manufacturing them into food by chlorophyll aided by sunlight.

The leafy green stuff that we think of as moss is the part of the plant bryologists call the gametophyte. There's a wonderful diversity of shapes in the leaves of different mosses: some are heart-shaped, some oval, some are tongue-shaped and some oblong, some have smooth margins and others are toothed, some come to a sharp point and others are quite blunt. This variety of foliage has led to some wonderful common names for mosses, such as knight's plume, bugs on a stick and humpbacked elves. Clustered at the tip of the leaves are small masses of cells, called gemmae. When dislodged from the leaves, these or other parts of the leaves can develop into new plants, in completely asexual reproduction.

The second distinct part of the plant is called the sporophyte, which grows when sexual reproduction takes place. Made up of a stalk with a tiny capsule at its top, the sporophyte develops partly as a parasite on the gametophyte. Some species appear never to produce sporophytes, depending entirely upon asexual reproduction, in which gemmae are cast off and carried by wind or water to new areas.

Sexual reproduction is far more intricate, far chancier, more dependent upon appropriate weather and a more beautiful process. Wilf Schofield describes the mechanics of the process in detail in his handbook, *Some Mosses of B.C.* In summary, the plant produces clusters of female sex organs, each shaped like a microscopic flask with an egg inside it, and surrounded with protective leaves. Equally protected, the male sex organs grow on different branches. When the plants are sufficiently moist, the thin walls of the male sex organs burst and sperms are released into nearby drops of water where, says Schofield, each sperm "swims about by means of two terminal hairs like flagella that are in constant motion."

Successful sperms manage to swim their way down the neck of the ruptured female sex organ and ultimately find an egg. Writes Schofield: "When a sperm touches the egg wall, it burrows through the wall and its nucleus unites with the nucleus of the egg." The fertilized cell commences cell division, and eventually grows to form the sporophyte. A capsule forms at the top of its stalk, housing a cluster of spores.

Just as the moss awaited moisture to facilitate fertilization, it now awaits a drying trend to aid in spore distribution. Dry weather first withers a protective sheath surrounding the capsule. Next the dried-out lid of the capsule pops off. Then, in some species, a ring of teeth around the opening serves as a weather-driven spore disperser. When the atmosphere is moist, the teeth bend into the capsule enclosing the spores. In doing so, they pick up a dusting of spores. Then, as the teeth dry out on warmer days, they rise upwards, in the process flicking the spores out of the capsule to be dispersed on the breeze.

Having landed, a spore, under the right moisture and temperature conditions, expands and splits its shell to produce a slender filament which creeps over the substrate like a living cobweb. This greenish web of threads attaches to the surface by means of colourless rhizoids, and in time buds of cell masses begin to differentiate into stems and leaves, thus initiating a new colony of moss.

* * * * *

These new colonies can take hold on a startling diversity of surfaces—everything from animal faeces to solid rock, depending upon species preference. Schofield says many mosses are very specific about what surface they'll colonize. There's one, for example, which will grow only on acidic rock in exposed locations. Others demand rocks rich in lime. Some will grow only on cliffs in certain parts of their range, and yet they'll colonize trees in other areas. Some prefer very dark places, and there's even a "luminous moss" which glows vivid yellow green in caverns and shady nooks of the rain forest. Another specialist grows only on animal droppings and disperses its spores by dusting the bodies of flies attracted by its foul smell.

Still other mosses, particularly many species of sphagnum moss, thrive in bogs and swamps, where there's abundant water and sunlight. Some species of sphagnum are aquatic, floating in quiet waters. These grow into a mat which absorbs substances from the water and renders it highly acidic. The acidic water inhibits the decay of organic materials, which begin accumulating in the bog. The quaking mat of sphagnum spreads across this accumulation and is eventually colonized by other plants which may be the precursors of a new forest community. Thus sphagnum plays a vital role in transforming open bog into forestland.

But the reverse can also occur. Schofield explains how sometimes sphagnum can grow outwards from a bog to colonize and transform adjacent forest. "The whole population acts as an immense absorptive sponge, water moving from the wet bog outwards to the drier perimeter." In sufficient quantities, this translocated water can drown the trees, which in turn increases available light and spurs new sphagnum growth. It's sphagnum's incredible absorptive capacity that makes it useful as a soil conditioner in gardens.

At the opposite end of the moss ecology spectrum, some species do best on exposed rock faces where they're subject to being completely dried out by the sun. These colonizers can play a key role in the centuries-long transformation of bare rock into a living forest. Clinging to the rock face by its anchoring rhizoids, the moss traps moisture and in the decay of its old growth produces certain acids. Water and acid unite to begin a chemical breakdown of the rock surface. Eventually small pockets of soil are produced, sufficient for other colonizing plants to get a toehold, and initiating a successional process which may culminate in enormous trees growing on what was once bare rock.

* * * * *

It's possible that just such a scenario unfolded on Haida Gwaii many thousands of years ago. During the last ice age, the mighty Cordilleran ice sheet covered virtually all of British Columbia, in places nearly a kilometre high. Retreating about ten thousand years ago, the glaciers left behind a landscape scraped bare of earth and of life. But scientists now think that certain small areas may have been spared the glaciers' icy advances. It's believed that along the fjords and headlands of the Queen Charlottes' western coasts, small refugia served as safe havens for certain plants—and perhaps for other life forms as well—holdouts through the long centuries of glaciation.

In time, as the ice sheet withdrew, these hardy survivors of the refugia set out to colonize the scraped rock and rubble. Today there can be found growing on the islands a number of plants completely distinct from any others on the planet. Botanists call them endemics. As well, the Charlottes are home to several disjunct species—plants found only in one or two other exotic and distant places. Some of these rare species are alpine and subalpine flowers, but the greatest number of them belong to the moss family, ancient throwbacks that escaped the glaciers and are, in Wilf Schofield's lovely description, "surviving out of their time."

A half-dozen true endemics have been identified—mosses that grow on the islands and nowhere else on earth. "An amateur would never see them," Schofield says. But three of them, recently described by colleagues, have been named after him, an honour the professor winces at a bit. As well, there are about a dozen mosses and liverworts that grow only in the Charlottes and one or two other places. For example, one liverwort grows only here and in Bhutan, another only in the Charlottes, Japan and New Guinea. And the list goes on. Of tremendous botanical importance, these plants have failed to generate any show-biz furore. "If they were animals," Schofield says, "it would be like finding a native monkey in the Charlottes; people would be extremely excited!" As it is, the excitement is confined to academics like himself who are painstakingly piecing together the true story of "the Canadian Galapagos."

* * * * *

Meanwhile, the average householder is painstakingly trying to get all that confounded moss out of the lawn. There are about a dozen moss species that become weedy in lawns, includng one exotic that appears to have been introduced from Europe. Wilf Schofield warns that some of these weedy mosses are capable of taking over "immense tracts" of lawn. The moss killers sold at gardening supply stores are inorganic herbicides, usually containing ferrous sulphate and ammonium sulphate. Gardening books tell us that lawns being overtaken by moss are typically suffering from some combination of poor drainage, lack of fertility and acidic soil. Aeration of the lawn along with regular applications of compost and lime will go a long way towards correcting the conditions that favour moss.

At home our attitude has always been: if moss grows in an area more willingly than grass, let it be moss. Now I read that moss is being touted as the "trendy ground cover of the nineties!" Elvin McDonald, a horticulturalist, author and secretary of the American Horticultural Society, is quoted as identifying a trend away from closely mowed grass lawns, "reducing the size of crew-cut perfection." Instead, "moss lawns" are being promoted. They're difficult to establish, are susceptible to invasion by weeds and require frequent watering; but the glory of moss lawns is that you can leave the lawnmower in the garage—they never need cutting!

Two very popular ground covers—green Irish moss and golden-green Scottish moss—form dense, compact mosslike masses of slender leaves and stems. Thriving in full sun or partial shade, these perennial plants are really not true mosses, but they do make a lovely ground cover, and we have them both growing in our gardens.

Elvin McDonald points out that the idea of a moss lawn comes from Japanese landscape gardening, in which mosses are carefully cultivated to simulate nature. Wilf Schofield writes that mosses can contribute considerably to a garden's beauty, particularly in covering boulders or in shaded and humid areas. However, the appropriate moss must be selected for various conditions and, once planted, must be tended as carefully as any other garden specimen. A mixture of moss and grass, for example, shows neither plant to good advantage and can look plain unsightly.

* * * * *

Back up on the Queen Charlottes, on our second day, we bailed out of a helicopter and scrambled crouching through long grasses dancing in the rotor's wind. Within moments the chopper lifted off the small mountain ledge where it had dropped us, and swooped away across forest lands to the east. We zippered our jackets against a chill wind and looked about us.

Shouldering up on our left was the craggy peak of Moresby Mountain

—at eleven hundred metres, the highest point on the archipelago. Fingers of green alpine heath stretched up among the summit's rocky outcrops. All around us, herb meadows formed an amphitheatre of vivid green splashed with white, yellow and purple wildflowers, which sloped perhaps two hundred metres down to a small lake. To the west, wisps of mist floated across the ridges of the outer mountains. The breeze carried a faint scent of crushed grasses and herbs. In the perfect stillness of the place we could hear, some-where far below, the clear sound of water splashing over rock. "It's like creation," one of us said.

Murmuring assent, we shouldered our packs and set off across the steep mountainside meadow in search of other rare mosses, which might hold fur-ther clues to how this extravagant beauty evolved and the unobtrusive but essential role mosses may have played in this fabulous unfolding.

Chapter Ten

MICE

From Steppes to Stars

*I*n 1967, UPI news service issued a breathless dispatch from Yugoslavia describing "waves of field mice gorging themselves across vast areas of Bosnian farmland." Feared to be carrying diseases, "of possible epidemic proportions," the mice were reported munching their way through croplands and invading homes, impervious to the stout resistance mounted by the embattled Bosnians. "Distraught peasants tried to stem the advancing menace," the report continued, "by rushing into their fields at night and killing the mice with hoes, shovels, rocks and other tools." But the fight was forlorn. One of the distraught peasants lamented, "The mice are not afraid of people. We stood there in the fields killing them, but they just kept coming."

I know the feeling—not of a mouse plague exactly, but of their inexhaustibility. Surrounded by woods and fields, our home is home as well to periodic plagues of mice. It's bad enough to have them scratching and cavorting in the walls at night, but it's downright unacceptable to spot them brazenly darting behind the woodstove in broad daylight!

I'm not even certain what kind of mice these intruders are. They might be native deer mice (*Peromyscus maniculatus*)—sometimes called white-footed or white-bellied mice—which range from Canada south through Mexico. Admirably adaptable, this species thrives everywhere from alpine meadows and boreal forests through woodlands, brushlands and grasslands to southern deserts and semitropical areas. And they're prolific, usually outnumbering any other mammal in their territory. That part certainly describes our kitchen, and deer mice are the one native species known to regularly invade buildings.

Then again, our uninvited guests could be harvest mice (*Reithrodontomys megalotus*), called American or western harvest mice, which resemble house mice in most details. Ranging from southern B.C. to Mexico and as far east as Indiana, the harvest mouse too is a tremendously adaptable character, thriving in everything from salt marshes to ancient rain forests, from below sea level to above the timber line.

These natives are two of over one thousand species of rodents found worldwide, comprising by far the largest family of mammals on earth. One of the rodent families, known as Muridae, includes rats and mice as well as hamsters, voles, lemmings and gerbils. Other natives, such as jumping mice and pocket mice, belong to entirely separate families. Of the sixteen murid subfamilies, one encompasses the so-called New World mice and rats, including deer and harvest mice. Also included is a white-footed mouse (*Peromyscus leucopus*) which closely resembles the deer mouse, and is found throughout the eastern and midwestern states. The northern grasshopper mouse (*Onychomys leucogaster*) is widespread through the American west and extends its range northwards into the southern prairie provinces. Another murid subfamily, Murinae, includes Old World rats and mice, and in this group we find the most familiar mouse of all—the common house mouse (*Mus musculus*).

One old story tells how the devil first created the mouse aboard Noah's ark; another has it that the devil himself stowed away on the ark in the form of a mouse, and soon mischievously set to gnawing a hole in its hull. Witches too have been blamed for creating mice, particularly in some of the grim folklore of northern Europe. In one version, a witch would take a scrap of cloth and fashion it into a little mouse, saying, "run along and come back," whereupon the mouse would spring to life and scurry away. Other witches in northern Germany used a different formula: while brewing magic herbs in a cauldron, the witch would chant, "Maus, Maus, heraus in Teufels Namen," and mice would leap magically from the pot!

Science predicates a different origin for the common house mouse. It's believed to derive from a wild species inhabiting the steppes and semideserts of Russian Turkestan. Gradually adapting to the development of human grain-growing in that region, the domestic mouse then journeyed with humans along the ancient caravan routes to the eastern Mediterranean, and from there through north Africa and into Spain. Eventually this race spread through western Europe—there's a separate subspecies east of the Elbe River—and scurried aboard boats setting sail for the New World.

As much as I favour the work of witches, I've had first-hand experience with the migrant mouse explanation. Camping on a mouse-infested stretch of shoreline in Baja one winter, we inadvertently adopted a mouse which scratched and gnawed away every night in our van, but defied discovery by day. We were half-way back to Canada (and half-mad to boot!) before we succeeded in enticing it out of the van and releasing it—with who knows what dire environmental consequences—fifteen hundred kilometres from its home!

* * * * *

They call *Mus musculus* a commensal species, literally meaning "eating at the same table," and in animals and plants referring to one that lives as a tenant of another species and shares its food. Mice are our tenants, and they've followed us everywhere, even to forbidding places like Antarctica where they thrive in heated buildings. No matter where we go, it seems, like it or not, we are stuck with *Mus musculus*.

Over the centuries our relationship with mice has ranged from adoration to rage. In the folklore of many ancient peoples the soul was believed to leave a dying person's body, slipping out of the mouth in the form of an animal, often a mouse or rat. A red mouse was said to betoken a pure soul, whereas a black mouse signified a corrupt soul. Sometimes a person would only be sleeping while its soul went wandering in the form of a mouse; if the mouse did not return, the sleeper would die, and it was thought dangerous to awaken a sleeper lest the soul be absent at the time.

Tame mice were often used in ancient religious cults. On the island of Tenedos in the eastern Aegean, for example, the Greeks erected a temple to Apollo about 1500 B.C. to celebrate a great military victory. They credited this particular triumph to mice which, before the battle was joined, crippled the enemy's fighting capacity by gnawing through their bowstrings and the leather straps which held their shields. It was sort of a pre-emptive mice-strike. Forever grateful, Apollo's priests on Tenedos bred white mice for use in prophecy and for carrying messages to the gods.

In other parts of Europe, too, white mice were esteemed as omens of good fortune, and in both African and American Indian folklore the mouse is seen as a benign and helpful character. But more often mice were feared as creatures of vengeance and ill-omen, possessors of an evil eye. In the Old Testament, when the Philistines pilfered the Ark of the Covenant from the Israelites' encampment, they were punished with a plague of mice and of haemorrhoids. Greatly discomfited, they eventually sent the Ark back to the Israelites, and in atonement sent along five golden mice and five golden haemorrhoids!

Even worse was the fate of a certain Archbishop Hatto who was eaten alive by hordes of starving mice in a mediaeval watchtower on the Rhine River near Bingen. This was one of several legends involving the infamous Mouse Tower, and most of them saw a heartless miser torn to death by rodents. Another old German tradition tells how mice will flee a house in which someone is about to die, the same way rats desert a sinking ship. For Greeks, mouse holes gnawed in flour sacks or clothing were an omen of hunger or death. In France the appearance of many mice was taken as a sure sign of imminent war.

As well, the mouse has contributed generously to various of our axioms and adages. "Poor as a church mouse" and "when the cat's away the mice will play" remain familiar expressions. Less frequently heard these days are old sayings such as, "it's a bold mouse that nestles in the cat's ear" and "the mouse that has only one hole is quickly taken." Not that long ago, mouse was common slang for a black eye, and a little bit farther back it was a common term of endearment—one of those diminutives men enjoy applying to women. Thus a sixteenth-century English writer named Warner wrote:

"God bless you, mouse," the bridegroom said,
And smakt her on the lips.

* * * * *

Small, poor and timid mice may be, but they're great athletes—excelling at running, jumping, climbing and swimming—and they're wonderful contortionists. A spokesperson for the U.S. National Pest Control Association recently called them, "the Houdinis of the animal world," because they can squeeze their pliable bodies through a hole no wider than your thumb. One day I surprised a mouse in our privy, and it disappeared instantly through a crack scarcely two centimetres wide.

Legendary for their breeding prowess, mice in the temperate southwest coast climate continue breeding throughout the year. Through most of their range deer mice have a breeding season from March through October. A female deer mouse normally produces three or four litters annually, with an average litter size of four. The house mouse also produces several litters each year, litter sizes averaging between four and seven, with one and twelve being the extremes. Offspring reach sexual maturity in about thirty-five days from birth. So in one year the progeny from a single pair of mice could be somewhere about twenty-five hundred.

Social animals living in small clans, house mice have evolved a form of "birth control" to limit population densities. Many of the females, especially the younger ones, simply don't become fertile. The ovaries remain inactive, the uterus thin and the vagina closed. Researchers believe stress connected with overcrowding may affect hormone secretion in the females, bringing about this population control.

Some house mice live outdoors, at least part of the year, and there are completely feral populations, particularly in dry grassland and farmland areas. But in forested, cold or wet places, they like to get indoors, especially for winter. Deer mice may also take advantage of the warmer winter quarters offered by buildings. At our house we find there's usually something of a fall migration to indoor winter quarters. There they can accommodate themselves to the tiniest compartments, with up to forty relatives crowded into communal nests. Noise doesn't seem to bother them—their hearing is pitched to very high-frequency sounds, so you can have your favourite heavy metal album thundering away at full volume, and for all the mice care you could be listening to Chopin. And they can thrive in frigid conditions too, so long as there's food available—mice have been discovered nesting happily inside the carcasses of frozen meat hanging in refrigerated storage chambers.

Mostly nocturnal, many mice are also active by day, and they're great power nappers, having been observed to alternate between sleeping and waking activity as many as twenty times within a twenty-four hour span. They prefer to live as close as possible to their food source, and sometimes a house mouse's entire territory may only cover a few square metres. Studies in Texas showed that western harvest mice have a range of about 0.2 hectares. Deer mice are also sedentary, with a range of about one hectare, with the most adventurous members travelling about a thousand metres from home.

Deer mice show very little inclination to fight over territory, and they'll sometimes intermingle with other species, particularly in winter nests. But according to Konrad Lorenz, house mice will defend their bit of territory with tremendous ferocity. In his classic *On Aggression*, Lorenz describes the experience of rodent researcher J. Eibl-Eibesfeldt:

> The house mice, which lived free in his hut, were regularly fed by him, and he moved about quietly and carefully so that they were soon tame enough for him to make observations at close quarters. One day he opened a large container in which he had bred a number of big, wild-coloured laboratory mice, not too different from the wild form. As soon as these mice dared to leave the cage and run about in the room, they were attacked furiously by the resident wild mice, and only after hard fighting did they manage to regain the safety of their prison, which they defended success-fully against invasion by the wild mice.

* * * * *

Less successful against invasion was a faint-hearted railway signalman in Southbourne, England, who a few years ago found his signal tower invaded by hordes of mice. Plainly no iron-willed Thatcherite, the myophobic (mouse-fearing) signalman shamelessly abandoned his post, leaving rush-hour trains stranded in chaos along the line.

Periodic plagues of mice seem to have carried right on from the days of the haemorrhoid-harried Philistines. Australia particularly seems to specialize in plagues of all sorts, alternating between toads, rabbits, mice and other introduced species. In 1970 a Reuters dispatch from Melbourne reported that "millions of ravenous mice are carving a five hundred-mile-wide trail of destruction across Australia's rich wheat belt in the country's worst plague in living memory." Farmers, said the report, were powerless against the invasion, and cats were "fleeing in terror!"

Ten years later another mouse plague swept across thirty-eight thousand square kilometres of South Australia, causing 6 million dollars in crop damage. "Marauding mice now reverting to cannibalism!" shrieked one headline. In the town of Woomelang, auto mechanic Bob Reid told the press: "Mice have eaten everything we had. . . . They wrecked our home. They're slowly wrecking our lives. I've seen mice chewing at the kids' ears at night, and eating at the corners of their mouths. When you lie in bed at night, you feel them run across your face. . . . It drives you crazy."

While not up to these same horror-movie dimensions, North American plagues have been more than bad enough for some people. "Housewives in frenzy over invasion of mice" read a 1953 headline from Alberta—a province which has long boasted of its rat-free status. In the 1970s mice infestations were reported in New York, Cleveland, St. Paul and Seattle, including a 1976 infestation in the White House, though that one was attributable to eight escaped white mice. In 1985 Vancouver had its own mini-plague in one east-end apartment building. With "herds of mice" infesting the building, city health officers declared themselves "ready to step in." Tenants, meanwhile, stepped out, abandoning the building and refusing to pay rent until something was done.

It's a mystery why these plagues erupt suddenly, apparently out of nowhere. Plainly they're evidence of an ecosystem out of whack. Explanations often run to a lack of natural predators. For example, the Bosnian plague of '67 was eventually blamed on hunters who had wiped out local fox populations that would normally have kept mice numbers down. Besides domestic cats, mice are naturally preyed upon by snakes, hawks, owls, weasels, opossums, skunks, raccoons and foxes. Eliminate these, and a mouse plague may not be far behind. Peak populations are often followed by an equally dramatic crash in numbers as food supplies are exhausted.

* * * * *

A rodent is literally "one who gnaws away," the word deriving from the same Latin source as "corrode" and "erosion." All rodents have twin pairs of specialized incisors, in the front of the mouth, which continue growing throughout the animal's life. The upper and lower incisors are abraded against one another, honing them into a pair of very sharp and efficient chisels. These are what mice employ to gnaw your house down. The Structural Pest Control Association of B.C. has said that mice, not rats, are the primary structural pest in the south coast area. Besides gnawing through structural members, they do a great job of tearing out fibreglass insulation for nesting material. They'll nest in walls, boxes, drawers, cupboards and mattresses.

Sometimes, not content with this level of destruction, they'll burn the whole place down: fires are often attributed to mice chewing through electrical wiring in walls or gnawing at matches. I wasn't able to find any documented cases, but was assured by several people whose business it is to know that "it does happen."

Outdoors, mice can be just as much a problem, particularly in orchards. In wintertime hungry mice will take to gnawing the bark of young fruit trees, sometimes girdling and killing the tree. The problem is lessened by clearing away sheltering mulch, vegetation or sod from around the base of the tree. Some growers recommend spreading crushed stone, gravel or sand around the base. Others favour tilling the orchard, thereby exposing any rodents to hunting owls and hawks. In severe cases, fine-mesh hardware cloth is wrapped around the tree trunk in a cylinder and embedded firmly in the soil.

In *The Golden Bough*, J.G. Frazer described a more dramatic approach to exorcising orchard mice.

On the eve of Twelfth Day in Normandy, men, women, and children run wildly through the fields and orchards with lighted torches, which they wave about the branches and dash against the trunks of the fruit trees for the sake of burning the moss and driving away the moles and field-mice. They believe that the ceremony fulfills the double object of exorcising the vermin whose multiplication would be a real calamity, and of imparting fecundity to the trees, the fields and even the cattle; they imagine that the more the ceremony is prolonged, the greater will be the crop of fruit next autumn.

In certain forest areas of the Pacific Northwest, and elsewhere, deer mice are a major factor in survival and dispersal of conifer seeds. One B.C. Forest Service study found that up to 90 per cent of lodgepole pine seeds were destroyed within a few weeks by high densities of mice, voles and chipmunks. In such areas the study recommended spreading sunflower seeds or other foods to ensure survival of enough tree seeds to produce a uniform cover of seedlings.

As silly as it might seem, feeding sunflower seeds to wild mice is simply one more episode in a very long drama of human efforts to control or eliminate mice. An ancient Greek treatise on farming advised the husbandman who would rid his lands of mice to

> take a sheet of paper and write on it as follows: "I adjure you, ye mice here present, that ye neither injure me nor suffer another mouse to do so. I give you yonder field [here the field was specified]; but if ever I catch you here again, by the Mother of the Gods I will rend you in seven pieces." Write this, and stick the paper on an unhewn stone in the field before sunrise, taking care to keep the written side up.

Frazer describes another remedy, less dependent upon mouse literacy, but no less eccentric.

> Sometimes the desired object is supposed to be attained by treating with high distinction one or two chosen individuals of the obnoxious species, while the rest are pursued with relentless vigour. In the east Indian island of Bali, the mice which ravage the rice fields are caught in great numbers, and burned in the same way that corpses are burned. But two of the captured mice are allowed to live, and receive a little packet of white linen. Then the people bow down before them, as before gods, and let them go.

An acquaintance of mine who works for the East Timor Support Network tells me that the peasants of East Timor, before that lovely island suffered brutal repression by the invading armies of Indonesia, used to maintain small shrines at the corners of their fields where offerings of food were left for mice and other creatures, thereby ensuring that the crops were not destroyed.

Having done away with these old remedies, the world now waits to beat a path to the doorway of someone who can build a better mousetrap. At home I use the old standby snap-trap, which I rate as moderately successful. I don't think the traditional cheese is a particularly effective bait. One of the old-timers here put me on to raisins, and they seem to work well. Some people swear by peanut butter. No matter what the bait, mice seem to learn to avoid the traps after a few of their mates have paid the ultimate price. I move my traps around the house, and sometimes leave them baited but not set in order to break down this resistance.

For those who absolutely refuse to draw blood, there are live traps available, marketed under cute names like "Havahart." The idea here is release the trapped mice in the country. Since I already live in the country, the concept doesn't appeal to me all that much, nor am I entirely sanguine about the prospect of millions of soft-hearted city dwellers releasing their rodents on my driveway!

"Vermin control" people often advise using a combination of traps and poison. Health officials also recommend this combination, using anticoagulant rodenticides because the carcass dehydrates, thus causing no odour problem. With the old baits like strychnine, the carcass would start to rot and produce an unpleasant odour. Of course, pest-proofing one's home is the best approach, otherwise you could continue poisoning mice ad infinitum.

Mice have demonstrated that they can develop resistance to various poisons. A few years ago, Britain had an outbreak of so-called "Super Mice" which had developed an immunity to Warfarin, the main rodent control used for two decades. Lamented London housewife Mrs. Mary Cecil, "The mice seem to love the usual poison I put down for them. When they've finished it off, they hang about with a sort of wistful look on their faces as though they're pleading for more."

* * * * *

A surfeit of these toothy tenants is no fun at all—in a particularly compassionate phase one time, we let our populations burgeon unmolested, to the point where they'd sprint across our faces at night, their cold little feet barely touching our skin as they skipped across the bed. On the other hand, to be without mice entirely would be a worse calamity. Like everything else, they have their proper place in the web of life, and they perform innumerable useful functions. They help control other pests: one New Brunswick study found that deer mice and other small mammals had opened and destroyed 80 per cent of the cocoons of the destructive saw fly. Other studies, in New Mexico and Texas, demonstrated the importance of mice in destroying the pods of the high plains grasshopper. In turn, mice are a major food source for hawks, owls and other predators.

While it might be argued that mice have benefitted more than we have in our millenias-old relationship, it is unarguable that the mouse has also contributed directly to our well-being in many ways. The use of mice in medicine has a venerable tradition stretching back into antiquity. Examination of the stomach contents of children from pre-dynastic Egypt reveals that they were fed mice, probably for medicinal reasons. In different cultures cooked mice have been prescribed in the treatment of measles, smallpox and whooping cough. The Roman writer Pliny noted that the ashes of mice could be mixed with honey as a cure for earache and halitosis. The English and Pennsylvania Germans believed that feeding children fried mice or mouse pie would cure them of bed-wetting.

More recently, laboratory mice have contributed enormously to advances in medical and genetics research. The familiar white mice of pet shops and laboratories are inbred albino varieties of *Mus musculus*. Deer mice have also been used extensively in genetics studies because they adapt readily to life in the laboratory and reproduce regularly. Not long ago I saw a news item describing how a professor of dairy science in Wisconsin was milking mice to test the effect of new genetic combinations on protein levels in milk. Attempting to develop "designer milks" for specific dairy products, the professor admitted that milking mice takes considerable skill and a very small milking machine. He estimated that it would take about six thousand milking mice to match the daily production of a single Holstein cow.

Always ready for an adventure, mice even got into space before we did. In 1960, three of them—named Sally, Amy and Moe—blasted off from Cape Canaveral in the nose cone of an Atlas rocket. Their flight of eight thousand kilometres at an altitude of eleven hundred kilometres was the farthest from earth that any living creature was known to have gone and returned alive. And how appropriate that this persistent tenant should have become a space traveller before we ourselves did. It's almost certain that if and when great space stations of the future glide effortlessly through the galaxies, there'll be colonies of stowaway mice on board, gnawing away somewhere down in the crafts' electronic bowels, and keeping everyone in touch with our distant, earthbound agricultural ancestry.

Chapter Eleven

RAVEN

The Great Transformer

We were hiking across a sun-baked salt pan in California's Death Valley and no creature seemed to stir in the afternoon glare. Then we heard a muffled *kwark* from a clump of stunted tamarisk trees. Gently I pulled aside the fronds of a twisted limb, and there, perched on a low branch not two metres away sat a common raven—lustrous feathers, black almost to purple, with black feathered legs, powerful claws and a hooked bill, black and blunt. Unnerved by my sudden appearance, the big bird shifted from foot to foot like a schoolboy caught mid-prank, cocked its head to look me over with a bright and inquisitive eye, and muttered something unintelligible in raven tongue. I bowed slightly and withdrew.

This is Raven, perhaps the most intelligent bird on earth, an enigmatic and contradictory creature, often despised by humans, and yet also revered; in myth a supernatural messenger as well as a slippery trickster. Here happily adapted to life below sea level in the arid oven of Death Valley, where daytime temperatures regularly soar well over thirty-eight degrees Celsius, I've seen ravens equally at home on soggy Haida Gwaii (the Queen Charlotte Islands), and gliding high above the Rocky Mountains. They roam the lands of the northern hemisphere from Iceland to North Africa, from Ellesmere Island to Afghanistan.

I know old Raven well (or think I do); a pair nest secretly somewhere among the dark conifers behind our house. Often they come flapping through the clearing with that distinctive whush, whush, whush of wings and a range of croaking cries. When a fresh summer wind blows from the northwest, they come out to play, gliding and rolling, banking against the breeze. Sometimes one will perch atop an old snag and call away to a distant companion. As payment for these obscure messages, they've helped themselves to duck and chicken eggs and hatchlings as well as delicacies from our compost heaps. Thieves and sneaks, if I catch them in the yard, they seem to laugh as they hop, twice, three times up into the air and are gone on strong wing strokes.

For the native peoples of the West Coast, Raven is a powerful mythical character—creator of the world and of human beings—but often foolish, always a bit devious and frequently greedy. Raven has a knack for mischief: making love to married women, getting caught and beaten up by outraged husbands and thrown into latrines or swallowed by whales. But he'd always turn up again like a bad penny, always re-emerge as the Great Transformer, accomplished practitioner of magic arts, a figure both laughable and tragic.

* * * * *

To the ornithologist the common raven is *Corvus corax*, the largest member of the family Corvidae, which also includes jays, magpies and crows. One of Raven's nicknames in England was Great Corbie-crow, and in various places both ravens and crows were called corbies. Though the two are closely related, ravens are larger than crows, more solitary and far less comfortable around humans. Corvids possess the largest brains relative to body size of all birds, and are generally believed to be the most intelligent avian family. Writing in *Natural History* magazine, zoologist Bernd Heinrich says, "The raven is, to my mind, the ultimate corvid in size, intelligence and inaccessibility."

Ravens are remarkably opportunistic, inventive and adaptable. Masters of mimicry, they are also capable of imitating the behaviour of other birds. And they love to fool around. Researchers believe that ravens have what may be the most complex forms of play of all birds, particularly chasing play; this produces learning interactions with the environment that enhance their ability to adapt to a wide variety of habitats.

The backwoods are full of tales of ravens performing tasks apparently too sophisticated for what we vain humans have come to deride as "bird brains." One of my favourite stories was recorded in 1950 by a chap named Alex Woods. It seems that a particular raven took to feeding on a wildcat carcass left on the ice in front of Woods's cabin in the B.C. interior. One day Woods noticed that the bird had rolled over on the ice as though it was dead, but soon got up and resumed eating. Intrigued, he observed that this performance was repeated whenever other ravens came into sight. The well-fed raven, Woods concluded, was playing dead to signal to the others that the carcass was poisoned, so that they wouldn't join in eating it.

The scheme came undone one day when the actor got up too soon and was spotted by the flock. They all alighted on the ice and surrounded the culprit. "First one would caw and croak and point toward the cunning raven," Woods recorded, "then another would do the same, then the whole lot would caw and croak together at the offender."

Entirely different raven behaviour shows up in Frank Craighead's book, *The Track of the Grizzly*, in which a raven, after spotting a carcass, "took off, and before long there was a growing assemblage of the large black birds. . . . It seemed that first one raven, then another, had communicated to the raven community that food was available." Bernd Heinrich became intrigued with raven gatherings in the Maine woods, and spent the best part of four winters observing them to find out if flocks at a carcass were, "the result of ravens actively recruiting their fellows, inviting them to the feast."

Banding studies and careful observation convinced him that ravens were intentionally giving a cry associated with food that was then passed along a vocal relay system and resulted in many birds arriving to feed at a carcass. But why? Why not keep the feast to oneself, the way Woods's selfish trickster did? Heinrich concluded that the ravens were not being magnanimous—far from it. The feeding crowds were primarily young birds, not yet mated, which wander for hundreds of kilometres. Mated adult birds, by contrast, remain in much the same area year after year. If a juvenile vagrant happens upon some food, it may be chased off by a dominant adult. Studies have shown that dominant ravens will intimidate subordinate ones by raising their feathers and strutting; young birds respond by tucking their heads down and sometimes even rolling over like a submissive dog. The call for others to come join the feast is, says Heinrich, a clever plot by the vagrants called "swamping"—gaining access to food by inviting in so many other birds the dominant resident adults can no longer defend it. Heinrich speculates that the gathering of juveniles may also serve as "a travelling 'disco' where ravens feed, evaluate one another as potential partners, and choose lifetime mates."

* * * * *

Ravens are great talkers and have a vocabulary whose complexities still baffle researchers. The most common call has been described as, "a hoarse, far-carrying, rather wooden croak or kwawk." As well, researchers have identified a large number of individual sounds for specific situations: threats, appeals for help, appeasement, requests to join in nest building, and so on.

Understanding the vocabulary becomes complicated because there are numerous individual variations in the calls. They are sometimes used in situations which seem to change their meaning, and copied sounds can be used instead of, or along with, specific innate calls. Experts in raven talk believe that the meaning of specific calls may be modified by changes in pitch, volume, length and sequence. And the meaning may be further elaborated by physical displays accompanying the calls—for example, the "let's have a chase!" call, *kuk kuk*, is often dramatized by back flips, barrel rolls and somersaults.

Raven's ability to know what's going on and to pass information along has given it the status of a cosmic gossip in the legends of many peoples. In Norse mythology, mighty Odin had two ravens perched upon his shoulders—one called Hugin (Mind) and the other Munin (Memory). Each morning the two would fly away to discover what was afoot in the world, and return each evening to whisper the news in Odin's ears. The Irish have an old phrase, "the Raven's Knowledge," meaning to see and know everything. Weather forecasting became a raven specialty, noted by Greek and Roman writers and even as scientific an observer as Francis Bacon, who wrote: "Ravens when they croak continuously denote wind: but if the croaking is interrupted, stifled or at longer intervals they show rain." In Christian mythology, the raven brings succour as well as information to the sainted—iconography shows both the prophet Elijah and St. Paul the hermit being fed by ravens; Saint Benedict has a raven at his feet, and St. Oswald has one perched upon his hand.

Raven's reputation as a far-sighted messenger led to its use, at least in myth, by navigators as a "shore-sighting" bird. In a Norse saga, the navigator Floki was aided in finding Iceland by three shore-sighting ravens. The ancient Babylonian *Epic of Gilgamesh* tells of a raven released after the flood, except it fails to return to the ark. Similarly in the biblical flood myth, Noah releases a raven to find land, but instead the bird wanders off to feed on drowned corpses. Noah, so the story goes, then convinced Jehovah to punish ravens for their perfidy by henceforth colouring them black where they had always been white. The Roman writer Ovid tells another version of how Raven's mischief-making caused it to be changed from white to black: Raven gossiped to Apollo that a Thessalian nymph whom the god loved passionately was being unfaithful to him. In rage, Apollo killed the nymph with a dart and, in Ovid's words:

> He blacked the Raven o'er,
> And bid him prate in his white plumes no more.

People who have raised pet ravens confirm these tales of the bird's uncanny intelligence and ability to communicate by sound and physical gesture. Konrad Lorenz, who was fascinated with raven communication, tells of how a pet raven learned to substitute its name, Roah, for the innate raven call *crackcrackcrackcrack*, which means something like "Come fly with me!" Ornithologist Derek Goodwin writes: "The eyes are an important indicator of mood and intention in the raven as in man. Captive ravens often respond at once to a friendly, intimidating, or aggressive look with appropriate calls or gestures . . . even people not at all familiar with ravens can tell at once whether the bird's look is hostile, appeasing or loving."

A Victoria, B.C., resident named Frank Beebe wrote about the raven his family raised: "I think the most remarkable thing about the raven was his ability to recognize individuals: individual people, dogs, cats, falcons—anything. He distrusted strangers in any form." And Beebe tells wonderful tales of the raven's endless capacity for mischievous fun. "For the cat he had nothing but contempt. If he yelled at it, it would run every time, so he made a game of chasing it, or better, slipping up on it very stealthily when it was asleep and pulling its tail, or yelling in its ear, then flying and diving after it as it fled. Life was a game to be lived dangerously.''

* * * * *

In the wild, the game begins with an early springtime courtship of spectacular aeronautics and elaborate displays and vocalizing. We watch these grand theatricals, like airborne operas, around our place every spring. Then, suddenly, these reckless public displays give way to an almost furtive secrecy. Mated for life, breeding pairs are almost obsessively cautious about their nesting sites, preferring high cliffs or old-growth trees in a dense forest. But in this too they show ready adaptability: they'll substitute telephone poles or abandoned buildings when natural nesting sites are scarce. Using sticks, as well as found materials, such as scraps of wire and the rib bones of sheep, in the nest structure, they'll complement the fine inner lining of fur and hair with wool or other fabrics teased into fluffy strands.

The female incubates three to six eggs which hatch in about twenty days. Both parents feed and tend the hatchlings until they fledge in about six weeks. Folklore has it that ravens make poor parents, but research has proven the reverse to be true. They meticulously prepare food for the nestlings, feeding them preferentially with a food-offering call. They carry water in their throats to the young (on hot days the female will soak her underfeathers and cover her offspring) and they are fastidious nest cleaners. If the young survive the fledgling stage, when they are vulnerable to predation, ravens normally enjoy long lives. The oldest recorded captive raven lived for twenty-nine years, while the record for longevity in the wild—based on bird band data—is thirteen years, four months.

The raven is a generalist, a highly intelligent opportunist who has adapted to a wide variety of habitats and climatic extremes. Even in the Arctic, ravens feel no need to flee southwards with the onset of winter. A flock can live contentedly through the long and bitterly cold season by feeding on a whale carcass, or on seals which are trapped and pecked to death on the ice, or by foraging at the garbage dumps of Arctic settlements. Ravens have long mastered the art of feeding off humans. In the near east they were regular camp followers, trailing after nomadic herdsmen to forage scraps. They've been observed following hunters in the expectation of a feast of entrails, and are said to shadow hunting cougars for the same reason.

Bernd Heinrich mentions the story of an Alaskan trapper who was led to a sleeping moose by a raven giving him the "come follow me" display. "Many natives," he writes, "believe that ravens regularly show wolves, as well as human hunters, where caribou are located." Some anthropologists think that the absence of raven bones in the prehistoric camps of northern native peoples is an indication that the bird was seen as sacred and never killed. In a strange parallel, superstition in Cornwall, England, claims that King Arthur lives still in the form of a raven and therefore the birds are not killed there either.

But by far the largest body of legend, as well as Raven's reputation as a vile and sinister creature, springs from its eating of carrion. In his nineteenth-century classic, *British Birds*, F.O. Morris writes that the raven's "unearthly sepulchral voice and mourning garb" confirm the superstition that "in all countries, and in remote ages," raven was seen as "the harbinger of death, as gifted with the faculty of anticipating what to him might prove a feast, and with the same motive to be a 'camp follower' on the battlefield." The *Anglo-Saxon Chronicle* records that after a bloody battle at Brunnanburh in 937 A.D., the battered survivors "left behind them sharing the dead the black horn-beaked raven and the hoary white-tailed eagle eating the carrion."

Feeding on human corpses brought on Raven's reputation as a creature of ill omen, a "devil's bird" which, like the vulture, anticipates death and hovers close by when it is imminent. So Poe calls his midnight visitor a "grim, ungainly, ghastly, gaunt, and ominous bird of yore." Legend tells that the Roman orator Cicero was forewarned of his death by ravens, and on the day of his death one entered his bedchamber and pulled the sheets from his bed. In a similar vein, the poet John Gay wrote:

> The boding raven on her cottage sat,
> And with hoarse croakings warned us of our fate.

And Marlowe wrote of "the sad-presaging raven, that tolls the sick man's passport in her hollow beak." Bloody-minded Lady Macbeth rants:

> The raven himself is hoarse
> That croaks the fatal entrance of Duncan
> Under my battlements.

If battles were scarce, ravens would make themselves at home around places of execution. "Ravenstone" was originally the old stone gibbet of

Germany upon which ravens would perch awaiting the death of some poor wretch. They patrolled the lawns of the Tower of London when public hangings were common entertainment there. Rather perversely, a legend sprang up that if ever ravens abandoned the Tower, the British Kingdom would fall. For three hundred years superstitious Britons imported fledgling ravens; finally, in 1989, a raven chick was successfully hatched at the Tower.

In this century, ravens have learned to patrol highways where they devour roadside kills. Voracious as well as omnivorous, they'll cache what they cannot eat and are believed to have a prodigious capacity to remember where they've hidden food. Interestingly, the words "ravenous" and "ravening" (as in "ravening wolves") derive from French and mean "to devour voraciously," not from the Old English name for the bird, *hraefn*.

* * * * *

In many parts of North America, the raven's appetite has earned the enmity of ranchers. Lamb kill problems have been reported in the northwestern states, and in northwestern Ontario and Manitoba ranchers have had problems with ravens killing calves. But when I interviewed Wayne Weber, who works as a wildlife damage specialist for the B.C. Ministry of Agriculture, he told me, "I don't know anywhere else where ravens are as much a problem as in B.C."

Ravens are massing by the hundreds in certain areas of the province—a recent Christmas bird count tallied nine hundred of them near one small Interior town. Ever the opportunists, they're scavenging at municipal garbage dumps and then spreading out to reportedly play havoc with livestock and with nesting wildfowl and songbirds. Weber told me there'd been quite a few reports of attacks on calves in the Chilcotin ranching country, "as many as fifteen or twenty calves attacked on a single ranch." Small hog operators in the southeastern part of the province have had problems too. "The ravens will go right into the farrowing huts and kill newborn piglets," Weber said. "They've had up to three hundred piglets killed in one year."

Newborn lambs are the most vulnerable of all. On Saltspring Island, where ravens were accused of killing up to one hundred lambs a year, Weber's branch brought in an ornithologist to study the situation. During a four-month period, twenty-eight lambs and three adult ewes were reported killed by ravens. Twenty years ago, the birds were seldom seen on Saltspring and attacks on lambs were unheard-of. Once again, an overflowing garbage dump played a part in the equation—upwards of three hundred ravens were congregating there.

I've read that in Britain, where both sheep and corbies are plentiful, there is no predation problem. But a century ago, F.O. Morris wrote: "Live stock, as well, however, it stows away; weak sheep and lambs, as also poultry, it cruelly destroys; hence its own destruction by shepherds and others, and hence again its own consequent shyness and resort to some place of refuge."

As well, ravens are sometimes blamed for killing an animal because they're observed scavenging on a carcass which may have met death some other way. Experts say lack of information is hindering the search for solutions since relatively little is known about the general biology of ravens in many areas.

In response to ranchers' frustrations and some lurid newspaper accounts of raven attacks, B.C. authorities have legalized hunting of ravens—several regions now have an open season with a bag limit of five birds. However, shooting a raven is easier said than done—their use of sentries and their uncanny ability to recognize a gun from hundreds of metres away make them a far more challenging target than many of the so-called "game birds." One hunting enthusiast recently called the raven, "the ultimate sporting bird in B.C., a cunning mind on the other side of the gun!"

Weber advocates a more intelligent approach. "If we can deal with the problem without killing the birds, we prefer to do so," he told me, recommending dump clean-ups and indoor lambing where possible. "We're sensitive to the fact that ravens are important in native consciousness—we don't want to cause some of the same problems as the wolf kill has caused."

So there's old Raven again—provoking, causing mischief, stirring things up, and throwing our throwaway culture back in our faces. Meanwhile, researchers probe the subtleties of raven language, and ecologists point out the valuable role ravens have in cleaning up human garbage scattered across the North. What is this divine messenger, trickster and great transformer telling us now? As I listen to ravens shrieking, squawking and kwarking in the tall firs around our place, I wonder ...

Chapter Twelve

RACCOONS

Crafty Handiwork

*S*ome people think of raccoons as wonderfully intelligent creatures, affectionate, playful and downright adorable. To others they're vicious, devious and destructive vermin. Almost everyone's intrigued by them and no one disagrees that they're smart as a whip. Sterling North, a writer who cashed in on the coon craze by writing *Rascal*, an international bestseller that sold well over a million copies, described them as "perhaps the most intelligent, and certainly the most dextrous and adaptable mammal in North America, barring only man himself."

"There is a beast they call Aroughcun, much like a badger," wrote a sea captain named John Smith in 1612. Smith was attempting to spell a word of the Powhaten dialect of Algonquin. Other Algonquins called it Arakunem, and elsewhere observers recorded native names for this unknown New World creature as Arathkane or Arakun. The various names all mean, "the one that scratches with its hands," and today's common name derives from that precise description. To the natives with whom it shared the woodlands, Arakunem was a trickster who appeared in many stories, but not as creator or transformer the way major animal tricksters did. Nobody's fool, Arakunem was never caught out in his tricks nor made to suffer indignities as Coyote and Raven sometimes were.

With its fixation on categorization, European science failed to find as accurate a name for Arakunem as North American natives had done. Linnaeus first called the raccoon *Ursus lotor*, meaning "washing bear," and later changed it to *Procyon lotor*. *Lotor* means "the washer" and *Procyon* means "before the dog," referring to a group of stars which rise before the dog-star. For a long time naturalists believed that a raccoon will always wash its food before eating. Off on this tangent, the Germans called it *wachbaren* or *washbar*, the Spanish *ositos lavadores*, and the French *ratons laveurs*. It's now generally agreed that coons don't wash food, though captive animals may dunk food in water to simulate wild foraging in much the way a cat simulates hunting with a dead mouse. However, science was correct in connecting coons with bears, as they share a common ancestry and are closely related to pandas. Of the seven species of raccoon, most are tropical. The one familiar to North Americans ranges from Central America to southern Canada, with a number of regional subspecies.

* * * * *

Primarily a creature of woodlands, raccoons originally lived in forests and brush country, usually close to waterways. Highly independent and somewhat solitary, they hunt by night, camouflaged by their distinctive coats, moving noiselessly through the bush, slow but strong swimmers and adept climbers. By day they rest in the crooks and hollows of high trees, with a predilection for maples, elms and basswoods. Biologists say that about ten individuals can live in 2.5 square kilometres. In the wild they live for perhaps a dozen years. Though they look like roly-poly little bears or extraordinarily fat cats, they're really rather skinny under that dense, double-layered coat of fur; males weigh only about 9 kilograms and females are even lighter. However, the largest recorded specimen weighed in at 28 kilograms!

Their senses are extremely fine. One laboratory reported that raccoons have the keenest sense of hearing of any mammal they'd tested. Alert to danger, they'll cock their silver-tipped ears this way and that to pick up warning sounds or vibrations. Less reliant on scent, still their doglike glistening black noses can sniff out food where yours or mine might not. Their coal black, beady round eyes are very sharp, particularly at night, when they glow fiery red caught in a sudden beam of light.

But, as the natives knew in naming them, it is Arakunem's sense of touch that is most remarkable. Their forepaws are the deft and delicate instruments of a tactile maestro. With five long, slender fingers, the forefeet resemble hands and have a dexterity similar to a monkey's paw. Sterling North writes: "The raccoon's prehensile hands are endlessly busy feeling the shallows of lakes and streams for crayfish, hellgrammites, pollywogs, salamanders, min-nows and small shellfish." So adept are they in their explorations, writes North, the raccoon doesn't even bother to look, but rather gazes off into the moonlit night while its fingers dance among the shallows. Strong enough to pry open oyster shells, the black-skinned paws are also wonderfully soft. North continues: "Feeling the gentle hands of a raccoon upon your face is a rare experience. It's as though a blind person were seeking to understand your character through fingertips alone." Equally long and soft, the hind feet are not so supple. In a lovely turn of phrase, Joan Harris described a raccoon's trail as "like the imprints made by a tiny elf-child going barefoot."

Intelligent and sometimes gregarious, raccoons are accomplished communicators. North claims they are among the most highly articulate of all wild animals: "I would say that raccoons utter thirteen or fourteen calls and many nuances of each. There are tender little tremolos of various pitches which are almost songs, many meaningful notes of warning and the continuous family conversation which helps to keep the mother and her tribe together on their nightly foraging trips." Other coon sounds include hissing, purring, trilling and churring. Joan Harris writes of a coon she raised: "She could growl like a dog and when hungry made a sound half-way between a whine and a purr. Short, sharp barks and snapping jaws denoted anger. . . . [She emitted] a thin squeak at the top of her voice range . . . when alarmed or cornered.''

* * * * *

The raccoon's year begins in February with the onset of breeding season. Polygamous males roam the woods in search of females. Holed up in her den, a female will rebuff the advances of unsuitable suitors, but when the right one happens along, she'll accept him into her den for several weeks of sexual activity. Afterwards, he'll depart in search of other mates while she's left with a family to bear and raise.

After about sixty-three days, she gives birth, usually to four kittens. Born blind and weighing only about fifty grams, the tiny kits are covered with velvety short grey hair. In about two weeks their eyes open, and another eight weeks after that they're ready to venture out of the flea-infested den. Here they may get their first taste of the perilous world they're entering. They must descend as much as twenty-five metres down the trunk of the den tree, and in doing so expose themselves to perhaps their deadliest predator, the great horned owl. As the little kittens make their unsteady, head-first descent, an owl or hawk may swoop in silently and pluck one screaming off the trunk.

The mother will spend most of the summer teaching her brood how to survive in the wild, to hunt for food, to sense danger and to seek the safety of tall trees when menaced by hunting bobcat, cougar or wolf. Weaned after four months, the young remain with their mother until winter, and may den up with her as well. In temperate climates, raccoons remain active throughout the year, but in colder zones they find a warm den, again preferring a hole in a tall deciduous tree. They don't hibernate so much as sleep, piled perhaps two or three in a heap, surviving on a layer of accumulated fat. If the temperature gets up above freezing, they'll stagger out in search of food.

Food is the raccoon's principal preoccupation. Omnivorous to the edge of gluttony, it will eat whatever is available, often using its clever hands to pluck morsels from small hiding places. It will turn over rocks in search of snails, snakes or salamanders. Wading in creeks and ponds, it's quick and clever enough to snatch small fish or frogs. Seaboard raccoons will prowl a beach at low tide, finding mussels and clams and anything the tide has left. Omnivorous but fastidious, they won't touch spoiled or rotten food. They love eggs and will climb a tree to steal the eggs of nesting birds, or dig up turtle eggs from lakeside sandbars. They'll kill young muskrats and squirrels, as well as rabbits, voles and mice. In late summer, they'll raid the nests of termites, ants and hornets to eat their larvae, and the honey-soaked combs of wild bees are a favourite autumn treat. Slugs, grubs and worms, grasshoppers and crickets all disappear down the ravening gullet, along with nuts, berries and fruit. By autumn this feeding frenzy reaches an apotheosis as the coons gorge themselves on everything available, packing in as much as two kilograms a day, in order to store enough fat to get them through the winter.

* * * * *

It is raccoon's relentless appetite, along with its ready adaptability, that moves it from the category of cute "masked bandit" onto the vermin list. As humans have systematically destroyed their natural habitat, these brainy opportunists have quickly adapted and carved out a comfortable niche for themselves in the human-altered environment. If our forestry and agriculture eliminate their denning trees, they'll find a substitute den in a stump, hollow log or cave. In farmlands they'll take over a fox or woodchuck burrow. In cottage country they'll make themselves at home in cabin attics, woodsheds and boathouses.

They inhabit cities all across the continent, from Seattle to New York, living in parks and ravines, surviving by their wits, feeding from garbage cans and gardens, killing park birds and stealing their eggs. Within the vast and tree-lined tracts of suburbia they do even better, killing neighbourhood cats, disembowelling unwary dogs, plucking fish out of backyard pools, unhooking bird feeders and dropping them to the ground, prying garbage cans open, invading attics and even ripping open roofs, tearing up lawns in search of bulbs, stripping trees of apricots and cherries and decimating vegetable patches.

In farming country they can wreak havoc on fields of sweet corn or poultry farms. Like mink, raccoons can resort to killing frenzies of the most savage sort. Joan Harris writes: "Chicken farmers get up in the morning to find their birds decimated but not eaten, their heads pulled through the wire and chewed off." They'll commit the same sort of wanton destruction in corn fields, ripping off dozens of ears of sweet corn, taking a bite or two from each and scattering them all over the ground.

"Raccoons on rampage!" shout periodic headlines across the country. "Raccoons terrorize town!" In Vancouver, Toronto and New York complaints pour into city hall from harassed home-owners and pet owners. In one Vancouver neighbourhood, fifteen cats were killed by coons in a short and bloody episode, and officials began warning dog owners to keep their pets away from raccoons because some were carrying the canine distemper virus and salmonella. By 1987, the Vancouver SPCA was gassing over sixty raccoons a year, and trappers were catching several hundred a year, releasing them many kilometres from the city.

* * * * *

Live trapping and removal is the recommended solution for problem raccoons, but for the amateur trapper it can be a real battle of wits. To mix a metaphor, the coon's Achilles heel is its sweet tooth. Honey-soaked bread, marshmallows and peanut butter are all recommended as good baits. Eggs too can entice them into a trap. But no bait works if other good food is plentiful. Old-timers say that if you wait until the corn's ripe to start trapping coons, you'll wait a good long time and lose your corn to boot!

Over the years gardeners and farmers have tried all sorts of tricks to keep the masked bandit out of their corn patches. I've read of a farmer who claimed success by placing a transistor radio blaring at full volume in his corn field. Some people go with an electrified wire running down the corn rows about thirty centimetres off the ground. Blinking Christmas tree lights are said to work, though I have my doubts.

Some people try putting repellents—hot pepper, hydrated lime, garlic spray—right on the corn tassels. Beatrice Trum Hunter writes: "Out of sheer desperation, one gardener had a corn cage constructed for the sole purpose of growing corn and keeping raccoons out. The cage was made of 14-gauge galvanized wire mesh with steel posts set in concrete." Though this chain-link reaction worked, Hunter points out, "the obvious drawbacks of such a structure are the prohibitive cost and the lack of crop rotation."

Many gardeners, I suspect, reluctantly follow the course taken by my father whose succulent vegetable garden in suburban Toronto is constantly assailed by groundhogs, skunks, rabbits and raccoons from nearby Humber River conservation lands. The first three he manages to keep at bay with mesh fencing and live trapping. But raccoons are far better strategists than the others, and their repeated raids on his corn finally forced him to abandon growing it altogether.

* * * * *

But while the horticultural portion of the population labours mightily to keep coons out, there's an equally determined group that attracts coons with backyard feeding, or even keeps them as pets. Clean, smart, affectionate and easily trained, raccoons appear to make the perfect pet. Joan Harris believes that raccoons "seem charming in captivity because they are playful. The reason is simple: their intelligence and desire to be active drives them to play to relieve boredom and frustration." Far more curious than the proverbial cat, they get into all kinds of mischief and do remarkably clever things. They can learn to open a refrigerator, turn on a faucet to get a drink, or switch on a television and change stations until they find a program they like. Sterling North tells of one pet coon which, when instructed to shower, would go to an upstairs bathroom, turn on the shower and return dripping wet, to much applause and wonderment. Newspapers love to run pictures of raccoons flopped on someone's head or on a dog's back—I've even seen a photo of a raccoon sitting on a milking stool and suckling from a milking cow!

Given cutesy nicknames like Chucky and Candy, tame coons are never entirely tame. Although they're adorably playful as kits, Joan Harris writes: "Native ferocity of captive coons flares up when they are about twelve weeks old," and, "tame ones become fierce if they are left without handling, even for a day or two." Responsive to a familiar human voice, a pet raccoon will purr when praised and hiss when scolded, but a wild and unpredictable ferocity lurks beneath the domesticated surface. Joan Harris raised a female raccoon she called Vixen. "When upset she laid her ears flat against her skull, which gave her a mean and angry look. If she saw something alive she was not sure about, she backed away, baring her formidable teeth, the hair bristling along her spine, at the same time raising her backside higher than her head."

Camping along the California coast one time, we encountered a gang of raccoons—maybe seven or eight of them—which had found its ecological niche in alternately charming and terrorizing visitors to their campground. Some campers thought them awfully cute and tossed them scraps of barbecued meat. Others shrieked and scrambled into their campers. For the raccoons were brazen in their begging, and if begging proved fruitless were cunning in their thievery, and if theft failed became almost menacing in their little gang of ruffians.

My mind set on wilderness, I had no patience for their brashness and begging. Several times while I was preparing supper they hopped onto the picnic table when my back was turned. I shooed them off. They retreated. Advanced again. Again I shooed them away. They bared their teeth at me and hissed. All right, I said finally, you want a battle, let's have a battle! I grabbed my axe. I swung it viciously and whooped! I charged them; they scattered. I pursued them like a madman, growling and cursing. They fled before my wrath like fat little bears, their ridiculous ringed tails bouncing along behind them. After the skirmish I swaggered back to our campsite. We had no more visits from coons that evening, nor from any of our fellow campers either. More cowardly than contrite, the raccoons continued terrorizing other campers well into the night, and we drifted off to sleep to the distant shrieks from campers beset by those cunning beggars.

* * * * *

Natives traditionally hunted Arakunem for its hard-wearing fur and its fat, which they used in many ways. They applied it to joints swelled by rheumatism, and taught New Englanders that raccoon fat is, "excellent for bruises and aches." In his 1764 book, *New England's Prospect*, William Wood writes of the natives:

> Their smooth skins proceeded from the often anointing of their bodies with the oil of fishes, and the fat of eagles, with the fat of raccoons, which they hold in summer, the best antidote to keep the skin from blistering in the scorching sun; and it is their best armour against the musketoes, the surest expeller of the hairy excrement, and stops the pores of the body against the nipping winter's cold.

Settlers soon learned to imitate natives in eating coon meat, using the pelts for hats, and softening leather with the fat.

In later years, coon hunting became a "sport" bordering upon a passion, particularly in Appalachia and the Ozarks. Sterling North estimates that today about a million coons a year are killed for fun in what he calls "a sort of midnight madness." Specially bred and highly valued dogs—redbones, blueticks, black-and-tans—track their cunning prey in an attempt to tree it. A brilliant strategist, the coon employs a bag of tricks to elude the baying pursuit. Many reports tell of how a large raccoon can lure a hound into deep water and drown it by climbing on its back. Cornered by dogs, it will fight with deadly efficiency, keeping close to the ground and lunging for the hound's jugular. North says an adult raccoon can handle two or three dogs in close combat, and that a full pack is needed to kill it.

Other human interactions with raccoons have not been a whole lot more enlightened. The once-brisk trade in raccoon pelts has been a roller-coaster business, peaking in this century with the coonskin coat craze of the twenties flapper era. A pelt in those heady days could fetch eighteen dollars. Would-be entrepreneurs rushed to cash in by establishing fur farms. Within a few years, pelts were worth only five dollars and the get-rich-quick farms folded. One such addleheaded venture saw raccoons introduced to the Queen Charlotte Islands, where they have no natural predator, and where they now represent a significant threat to nesting birds.

On the other side of the continent, raccoons and foxes were purposefully introduced onto several small islands off the Massachusetts coast to control herring gull numbers. The gulls were supposedly posing a threat to aircraft using Boston's Logan Airport. The foxes and coons quickly eliminated about 90 per cent of the eggs and hatchlings and were declared by the planners, "an effective control measure."

The little island where I live has no raccoons, and there's a vigilant band of locals who'd be quick to form a posse if any misguided newcomer dared import a pet raccoon. My only other direct encounter with a raccoon was many years ago when I was a boy in Toronto, and our little gang of boys prowled the wild lands around a turgid little watercourse called Black Creek. I suspect the creek is lined with concrete and culverts now, but in the fifties its valley was as wild as the jungles of Sumatra.

On this particular day, our gang had taken a tremendous step, because one of our members had been given a new BB gun. Our old slingshots and peashooters seemed laughable next to this shiny new weapon, and we now stalked the woodlands with an enhanced sense of menace. Then we spotted a raccoon flopped in the crook of a tree. The Davey Crockett coonskin cap craze had barely been replaced by the hula hoop. We took aim. We fired. The first BB shot bounced off the coon's rump. We fired again. Same thing. We passed the gun around the whole gang. Every shot either missed or bounced off the coon, who was by now getting thoroughly annoyed. Reduced in our supply of shot and our sense of menace, we moved along. Eventually the boy who had been given the gun succeeded in shooting a tiny songbird perched in a bush. As the lovely little creature lay dead in his hand, I was shaken with revulsion and shame. I have never hunted again.

Chapter Thirteen

SPIDERS

The Sinister Spinners

*E*very once in a while I'm blasted out of deep sleep by a blood-curdling cry from my mate and her frantic thrashing about in the sheets. Even with the mists of sleep still clinging thick about me, there's no need to ask what's wrong. It has to be a spider. And it is. "Des, there's a huge spider in the bed!'' Sometimes, just often enough to sustain credibility, there actually is a hapless little spider scurrying about, and I perform my solemn duty of removing it to the out-of-doors.

Other times the culprit turns out to be a mistaken moth or mayfly or similar innocent. These too are removed, to palpable relief. The worst case scenario is when nothing shows itself. Then one's left to wonder: was this terror entirely imagined? A final search turns up nothing and we lie down again. With unbecoming ease she slips back into contented sleep while I'm left lying awake in the dark, senses alert, imagination active, nursing a growing conviction there's some great black hairy beast of a spider lurking with sinister intent somewhere among the bedclothes. This is the little-known phenomenon I call contact arachnophobia.

There is, after all, every good reason to be terrified of spiders. Menacing in appearance, exponential in numbers, devious in their devices, cunning mistresses of entrapment, venomous and merciless—who in their right mind has ever encountered an oversized spider and not felt a cold thrill of terror down the spine? The legs long and loosely articulated, covered with tiny black spikes, hairy and rapid in sinister advance. The body grotesque, with its bulging abdomen, its cephalothorax bristling with lethal jaws and palps and multiple eyes. Here is a superb killing machine, silent and murderous, which would if it could, and perhaps as its giant ancestors once did to our poor ancestors, bind you in its silken threads, immobilize you with a venomous bite and then slowly suck the living innards from your body.

I think I hear you sniggering behind your book. But consider the lumbering *Theraphosid* spider of Brazil. With a body about ten centimetres long and legs that can spread thirty centimetres, this is a consummate killer. In one experiment, a captive *Theraphosid* within the space of four days killed and sucked dry two large frogs and two highly venomous snakes. Then it rested, bloated and ominous, like Tolkien's Shelob in her lair. Picture yourself pursued by the swift and nimble running spiders of Central America, whose legs can easily cover a dinner plate. Think of blundering into the sticky web of a tropical orb weaver, a terrible web almost two metres in diameter hanging from six-metre guy wires. Or imagine waking up in Mexico one morning to find perched on your pillow the hairy tarantula whose nine-centimetre body and stout hairy legs can weigh fifty-five grams.

Scoff if you want. But put yourself for a moment in my position when, after an intense bout of contact arachnophobia, I perchance have to go to the toilet. At home we use a semidetached composting privy—

environmentally and politically correct, but to the average observer little more than an outhouse with a college degree. We know that in British Columbia there lives a species of black widow spider called the western widow (*Latrodectus hesperus*). We know that one of its favourite haunts is outbuildings. And we know that when outhouses were common, a choice hiding place for black widows was under toilet seats, and frequent reports were received of black widows inflicting painful bites upon exposed genitalia. Expose yourself, as I say, to my predicament for a moment and let's hear a bit more of your mockery!

* * * * *

Spiders have prowled the earth for something like 400 million years, the first of their kind being an aquatic scorpion. They belong to a class of predators called Arachnida which includes scorpions, pseudoscorpions, mites and ticks. Two special features distinguish spiders from the other arachnids: their ability to spin silk and the male's highly specialized oral sex apparatus. About thirty thousand species of spider have been identified worldwide, from the tiny to the grotesque, living almost everywhere, in scorching desert heat and polar cold, in the depths of caves and coal mines and on the wind-swept summits of mountains (a jumping spider was once spotted on Mount Everest at an elevation of 6,706 metres).

Wherever they live, they work as hunters or trappers, feeding off living prey. There is no vegetarian minority among spiders, no pacifist caucus; they are all killers of surpassing ingenuity and skill. Roughly they divide into two great camps: the hunters which stalk and capture prey directly, these comprising about 40 per cent of the species; and the larger group, trappers, which ensnare their victims in webs.

Tarantulas typify hunting spiders, where speed and strength are at a premium. The hunter's tactic is to pounce upon insect or other prey, either by stalking it or by lying in ambush and suddenly rushing out upon it. For the hunters, eyesight plays a more important role than it does for the poor-sighted web spinners. All spiders have eight eyes, usually set in twin rows of four, and in some of the hunters, like common jumping spiders, one pair of eyes is greatly enlarged, allowing them to hunt by day and detect insects a metre or more away. Once it has spotted an insect, the jumping spider approaches stealthily and then pounces. I've watched them leap with acrobatic skill across a crevasse and land sure-footedly on the tiniest of ledges.

Among the most fearsome of temperate zone hunters are the large, hairy wolf spiders, the so-called "northern tarantulas." They hunt in open fields and meadows, their markings providing subtle camouflage, their long legs swift in pursuit and efficient in seizing prey, which is wrestled into submission and then crushed in the wolf spider's powerful jaws. Many of these roving hunters are capable of tremendous bursts of speed, sprints covering fifty centimetres or more per second. Specialized hairy pads at the tips of their legs help them dash over difficult terrain. Some are even adapted to run across the surface of water: they'll wait in hiding, their legs touching the surface, alert to the telltale tremblings of an insect caught in the water, and then glide out to seize it. Big nurseryweb spiders—the largest species found in Canada—can even catch minnows or aquatic insects underwater.

Many of these hunters are experts in camouflage, as adept as a chameleon at changing colours or blending into backgrounds. Crab spiders, which resemble crabs by virtue of their pincerlike enlarged front legs and their habit of scuttling in all directions crab-wise, are especially good at changing their colouring to mimic the blossoms within which they hide, awaiting bees or other unsuspecting insects.

* * * * *

Our word spider derives from the Old English word *spinnan*, meaning "to spin," and though hunting spiders do not spin webs, they do, like all spiders, produce silk of different types, each for a specific purpose. Manufactured through special glands located in the creature's abdomen, and spun out in filaments through tiny tubes called spinnerets, the silk is pure protein and one of the most remarkable substances found in all of nature. It has a greater tensile strength than steel and an elasticity that allows some threads to stretch to twice their normal length before snapping. Tough and resilient, the silk is impervious to moisture, vibration and temperature—and it's totally recyclable: some spiders can eat a web and reprocess old silk into new within thirty minutes!

In their roving, hunters lay down a drag line of silk which trails behind them and which they attach periodically to the surface, so that if they leap from a precipice in pursuit of prey, the safety line allows them to descend without harm, and to re-ascend if necessary. But it is the web spinners, the 60 per cent of spider species that trap their prey in webs, who are the real mistresses of silk. This is the mythical Arachne, the most accomplished of spinners, creating space webs, sheet webs, cob webs, orb webs and funnel webs; webs that are no more than a few simple strands, and webs that are elaborate, complex tapestries; webs that last a single day, and webs that endure for the whole lifetime of their spinner.

"What a refinement of art for a mess of Flies!" exclaimed the French entomologist J. Henri Fabré, who spent half the nineteenth century observing and recording the behaviour of insects. "Nowhere in the whole animal kingdom has the need to eat inspired a more cunning industry." In *The Life of the Spider*, a book distilled from ten volumes of insect observations, Fabré tells how a large spider in his house provided entertainment for his whole family. "Big and little we stood amazed at her wealth of belly and her exuberant somersaults in the maze of quivering ropes; we admire the faultless geometry of the net as it gradually takes shape. All agleam in the lantern-light, the work becomes a fairy orb, which seems woven of moonbeams."

He's describing the work of an orb weaver. The most elaborate of all webs, an orb is generally a two-dimensional symmetrical spiral set on spokes and hung vertically from supporting structures. These are the webs we see in our gardens outlined like silver filigree with the dewdrops of autumn. The orb "glitters in the sun, looks as though it were knotted and gives the impression of a chaplet of atoms." Excited, the Frenchman takes a section of web and places it under his microscope.

> The sight is perfectly astounding. Those threads, on the borderland between the visible and the invisible, are very closely twisted and twine, similar to the gold cords of our officers' sword-knots. Moreover they are hollow. . . . the spiral thread is a capillary tube finer then any our physics will ever know. It is rolled into a twist so as to possess an elasticity that allows it, without breaking, to yield to the tugs of the captured prey: it holds a supply of sticky matter in reserve in its tube. . . . it is simply marvellous.

Exploring further, Fabré discovered that the framework of the orb web and the platform where the spinner rests is not sticky at all, but is constructed of "plain, straight, solid thread." But the filaments intended to catch insects—the "snaring" or "lime" threads—are coated with a sticky, viscous substance. How does the spider prevent its own hairy body from sticking to the web? The tips of its eight legs are ingeniously adapted to the task. Each ends in a tiny paired claw along with a single claw which hooks onto a strand of thread as the spider runs along it. Clusters of small, stiff hairs guide the thread on and off the swiftly moving hook. Frequently the spider grooms itself, cleaning and moistening each leg tip with its fangs.

Although orb weavers show tremendous ingenuity in constructing their webs—always adapting them to the supports from which they're strung, working primarily by instinct and touch to create masterpieces of combined straight lines—many are utterly incapable of the simplest patch job to repair a damaged web. Some get around the problem by reconstructing their webs, perhaps even on a daily basis, to maintain the stickiness. Others, notes Fabré, "reconstruct theirs only very seldom and use them even when extremely dilapidated. They go on hunting with shapeless rags. Before they bring themselves to weave a new web, the old one has to be ruined beyond recognition."

Resting at the centre of its ingenious snare or in a hiding-place at the edge, the poor-sighted weaver again depends upon its splendid sense of touch to know when the web has done its deadly work. If hiding at a distance, the spider stays in touch with the web through a signalling thread, which will alert it instantly if an insect blunders into the net. "Clutching her telephone wire with a toe," writes Fabré, "the Spider listens with her leg; she perceives the innermost vibrations; she distinguishes between the vibration proceeding from a prisoner and the mere shaking caused by the wind." Without the signal wire—which in some species may stretch three metres or more—the spider would remain oblivious to what is happening. In experiments, Fabré carefully snipped the signal wire and then placed struggling insects on the web, but the spider remained unaware of them.

Once alerted to vibrations in the web, the spider reacts instantly. If hiding off at a distance, it dashes to the centre of the web and plucks with its legs at the radial threads. More skilful than any harpist, it knows from the vibrations of the threads exactly where the insect is and how large it is. The spider goes directly to it. If the captive is small and harmless, the spider attacks quickly. "Facing her prisoner," writes Fabré, "the Spider contracts her abdomen slightly and touches the insect for a moment with the end of her spinnerets," then swathes the creature in a silken straight-jacket. For more dangerous prey, it turns its back and flings out thread in a battery of silk,

> firing a regular volley of ribbons and sheets, which a wide movement of the legs spreads fan-wise and flings over the entangled prisoner. . . . the most fiery prey is promptly mastered under this avalanche. In vain, the Mantis tries to open her saw-toothed arm guards; in vain, the Hornet makes

play with her dagger; in vain, the Beetle stiffens his leg and arches his back: a fresh wave of threads swoops down and paralyzes every effort.

Either before or after subduing the insect in silk, the spider bites the prey, injecting venom through its fangs and stunning the victim. The spider may then eat it on the spot, carry it back to the resting platform or leave it dangling in the web while attending to another victim. Spiders don't tear off chunks of food and swallow them; they inject the prey with digestive enzymes which liquefy the insect's body, allowing the spider to suck the innards out, leaving only a hard exoskeleton. "The mouth lingers, close applied, at the point originally bitten," Fabré writes. "It is a sort of continuous kiss."

*   *   *   *   *

While most spiders are venomous, only a few pose any kind of danger to humans, and even with these the dangers fail to match our fears. The most venomous of all is a large and antagonistic Brazilian brute of the genus *Phoneutria*, which has the nasty habit of getting inside clothing and shoes and then biting viciously when disturbed. Hundreds of such bites are reported annually. Extremely powerful, the neurotoxic venom injected with the bite quickly attacks the nervous system, resulting in acute pain followed by hallucinations, spasms and eventual paralysis of the respiratory system. Small children are particularly susceptible, and there are verified accounts of children under five years old dying horribly from *Phoneutria* bites.

A bite from the black widow, which lives in nearly all parts of the continental United States and southern Canada, produces many of the same symptoms—profuse sweating, cramps, disorientation and a tightening of the abdomen. It's like an attack of appendicitis. A widow's venom is many times more toxic than that of a rattlesnake, though of course the quantities are far less. Certain natives of the U.S. southwest used to combine the venom of both, and tip their arrowheads with the deadly mix to bring down big game.

A shy creature that hides in its tangled cobweb by day and hunts by night, the black widow will only bite a human when frightened or cornered. There are no reliable figures on fatalities from black widow bites, in part because spiders are often blamed for bites actually inflicted by other insects. One oft-quoted statistic is that between 1726 and 1943 there were 1,291 reported black widow bites in the U.S., half of these in California, and fifty-five fatalities. In other words, death from spider-bite is a long shot. Apart from a few of the heavyweight species—the fearsome tarantula, for example, has a bite equivalent to a wasp's sting that is painful for about an hour—most spiders don't have jaws strong enough to pierce human skin.

* * * * *

A lot of the bad press that spiders in general, and black widows in particular, receive derives from their mating rituals, which sometimes end in cannibalism. There seems to be something fundamentally offensive to us about having two creatures bond in the embrace of procreation and then have one eat the other. It flies in the face of our carefully cultivated notions of romantic attachment, making a cruel mockery of holy matrimony. The shocking realization that it is always the male who is devoured is even more intolerable in certain circles. No wonder the spider was once despised as a familiar of witches, hiding in the hood of a crone's cloak and whispering dark secrets in her ear!

Of course, if anyone is to be eaten after copulation, simple biology dictates that it not be the female, since that would defeat the procreative point of the exercise. The black widow, writes Adrian Forsyth, "is not entirely deserving of her name. In many cases she fails to eat her mate and he may live to mate again." But there are other spiders who practice what Forsyth calls "suicidal male monogamy." To understand this strange behaviour, we must look at the equally peculiar way spiders copulate.

The ritual begins with an adult male spinning a tiny patch of silk upon which he deposits, after tapping himself, a drop of seminal fluid. He reverses his field and siphons the fluid up into a pair of sensory organs called palps, located behind his jaws. Thus palpably prepared, he saunters out in search of sex. Locating a suitable female, he woos her with courtly behaviour. Fabré describes his approach:

The whipper-snapper [comes] to pay his respects to the portly giantess:
... he advances circumspectly, step by step. He stops some distance away, irresolute. Shall he go closer? Is this the right moment? No. The other lifts a limb and the scared visitor hurries down again.

Eventually, Fabré goes on,

perseverance spells success. The pair are now face to face, she motionless and grave, he all excitement.... With his legs, and especially his palpi, or feelers, he teases the buxom gossip, who answers with curious skips and bounds . . . [and] a number of back somersaults, like those of an acrobat on the trapeze. Having done this, she presents the under-part of her paunch to the dwarf and allows him to fumble at it a little with his feelers. Nothing more: it is done.

What he has done is transfer his seminal fluid by inserting his palps into a matching pair of apertures in the underside of the female's abdomen. In some species he must crawl beneath the larger female to accomplish his mission; in others he sprawls on her back and reaches over, first one side then the other, to insert each palp.

Most precarious of all, Adrian Forsyth describes the extreme case of an orb weaver called *Araneus pallidus*, in which "the size and mating antics of the male seem designed to place him directly on the fangs of the female." The little male, it seems, can only get his palps properly inserted if the female seizes his abdomen with her pincers. If she doesn't, he keeps slipping off without getting the job done. Trouble is, once she's got her fangs into him, she can't control herself and sets about feeding on him. "He is forced," writes Forsyth, "into suicidal monogamy by the very design of his body and behaviour."

Whether she eats her mate or not, the female has no further need of him. She holds his sperm in her own little sperm bank, called the spermathecae, where it can remain viable even through a whole winter. Most spiders will lay several batches of eggs, using a portion of the stored sperm to fertilize each batch as the eggs pass from her oviducts. Laid on a little carpet of special silk, the eggs are swathed in a silken cocoon. The females of some species guard their eggs until they hatch, and many web spinners hang the cocoons within their webs. Fabré describes the cocoon:

> Her nest, a marvel of gracefulness, is a satin bag shaped like a tiny pear . . . under the outer wrapper, which is as stout as our woven stuffs and moreover perfectly waterproof, is a russet eiderdown of exquisite delicacy, a silky fluff resembling driven smoke. Nowhere does Mother-love prepare a softer bed.

Some hunting species, including wolf spiders, attach the cocoons to their spinnerets and carry them everywhere they go. We often see them in the autumn, the spiders looking grotesquely swollen with their huge burdens. They'll fight ferociously to defend these egg sacs. One of the most attentive of all spider mothers, the wolf spider even carries her tiny spiderlings on her back for a few days after they've hatched.

Within days of hatching and their first moult, the spiderlings are able to feed themselves and spin their own silk. They quickly disperse to begin their solitary lives. In some species they simply walk off, but in others the little spiderlings disperse dramatically by "ballooning." Each climbs to some high point, perhaps a blade of grass or other plant, and spins out a first thread, an infinitesimally fine gossamer which, stretched to perhaps a metre in the breeze, lifts the spider into the air and off it sails. Ballooning spiderlings have been spotted by mariners eighty kilometres or more from shore.

The life these wind-spread balloonists live may be very short indeed. Most species of spider complete their life cycle within a year. The longest-lived are certain of the tarantulas that can survive up to twenty-five years. The black widow, by contrast, is one of the shortest-lived of all, and does not survive a full year.

"The spider has a bad name," Fabré conceded. "To most of us, she represents an odious, noxious animal, which everyone hastens to crush underfoot. Against this summary verdict the observer sets the beast's industry, its talent as a weaver, its wiliness in the chase, its tragic nuptials and other characteristics of great interest." Modern arachnologists now recognize that the spider is one of the dominant predators of any terrestrial ecosystem. For example, a census taken in an English meadow in late summer determined that each hectare contained about five million spiders! In such concentrations, experts say, spiders consume many times the number of insects eaten by birds. In the control of pest insects the much-maligned spider may in fact be humankind's most important ally. Some have gone so far as to speculate that human habitation of many regions of earth is only made possible by the countless billions of spiders also living there. I believe this. I repeat it to my mate when we're suffering a bedtime spider attack. And it helps, believe me, it helps.

Chapter Fourteen

STARLINGS

Murmurations and a Mirror

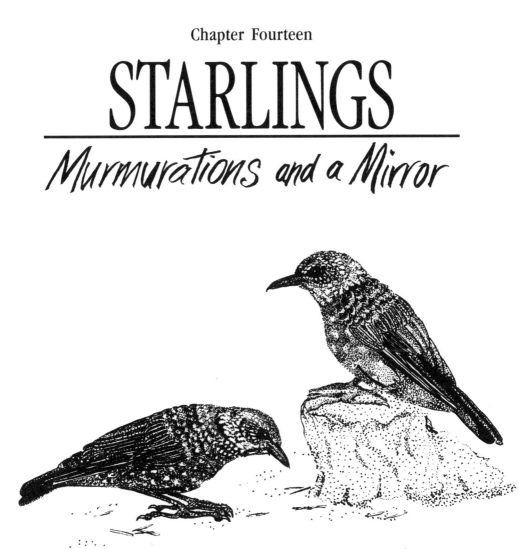

*T*he scene is a late summer evening in 1950s suburban Toronto.
Suffocatingly humid heat. Pat Boone
crooning about love letters on the
transistor radio. The awful tortured
boredom of pubescence. Then, suddenly, a commotion of slammed car
doors and muffled shouts nearby.

Within minutes, a tremendous volley of gunshots, a harsh screeching of birds, and a momentary hailstorm of buckshot bouncing off our roof. Yes, another great starling shoot was underway! Across the small schoolyard from our house, a couple of dozen men lined up along the street wielding shotguns, a policeman at their head. At a sign from the cop, the men would raise their guns, aim into the dense canopy of plane trees overhead and blast away in unison. Huge thunderheads of starlings would erupt from the trees, screaming in protest. A few of their number would flop to the ground and flap about pitifully, but most of the great black cloud would wheel in the sky and settle again in other trees farther down the road. Later on, the men would stand around talking and smoking cigarettes, shotguns slung easy over their shoulders, then climb back into their cars and drive away into the night.

By the time I left that old neighbourhood in 1960, the shooters had long since quit coming. But the starlings were still there, massed in their thousands in late summer evenings, roosting in the big plane trees, spattering the cars and sidewalks below them with white, and always talking, talking, talking. It's a scene that's repeated in cities and towns all across the country. The great starling invasion of North America is a story of spectacular biological success by a highly adaptable bird, as well as a cautionary tale on the folly of introducing exotic species.

If all the good citizens bespattered, awakened and harassed by massive starling flocks could vent their frustrations on one person, I'm certain they'd choose Eugene Schieffelin. A wealthy drug manufacturer in New York City, Schieffelin founded the American Acclimatization Society which had as its noble purpose the introduction into North America of every bird mentioned in the works of William Shakespeare. Unfortunately, in *Henry IV*, the bard has Henry Percy promise

> Nay I'll have a starling shall be taught
> To speak nothing but "Mortimer."

In fairness, the starling's reputation a century ago was not what it is today. In 1850, amateur ornithologist Rev. F.O. Morris, who lived in Yorkshire, gave this adjective-enriched assessment of starlings in his classic work, *British Birds*:

> They may be seen now sweeping off from their secure retreats in the grey old church tower, or the "cool grot" of the lonely cliff that overhangs the pebbled beach of the glorious ocean, and hurrying to the ploughed field or the farmyard, the quiet cow-fold and the pasturing herd, now

perching on an adjoining wall, and now on the back of a familiar sheep, and now whistling their quaint ditty from the house-top or a neighbouring tree.

Perhaps with similar visions of pastoral charm, Schieffelin imported some 140 European, or common, starlings (*Sturnus vulgaris*) in 1890 and '91 and released them in Central Park. Set loose in the New World, the population exploded and dispersed with astounding vigour. The Maritimes already had a thriving population by 1915. By the end of the First World War, they'd crossed the Great Lakes to southern Ontario, and within fifteen years were said to be the most numerous bird in many southern parts of that province. The U.S. Department of Agriculture confidently predicted in 1928 that the prairies would form an insurmountable barrier to starlings and halt their westward migration. The starlings paid no attention and were crossing the Rockies by the early 1940s. The first British Columbia sighting occurred at the little Okanagan town of Oliver in 1947, which should have served as a grim warning to the area's many cherry growers.

While the futile blasts of shotguns were exploding into the muggy evenings of my Toronto adolescence, westerners were beginning to see the first trickles of what would become a monstrous black tidal wave breaking over them. In 1950 the Oregon Audubon Society counted 43 starlings in the Portland area, far more than had ever been seen in the City of Roses before. In B.C., the Vancouver Natural History Society annual bird count tallied a paltry 5 starlings in '52. By '55 the number had shot up to 6,000. Ten years later it was 300,000 and biologists were warning that city officials had better "do something." By 1982 there were an estimated 4 million starlings roosting in the city, despite a brave pronouncement from the city Parks Board that "Vancouver is winning its battle with the dreaded starling."

* * * * *

I had my first hands-on dealings with what one local politician called "this plague of rats on the wing" at a small cottage in the Vancouver suburb of Richmond where we were visited by nocturnal rats. In the springtime, a pair of starlings nested in a small hole under the ramshackle building's eaves. Busy with burgeoning careers, we thought nothing about it until a foul odour began to permeate the house. Eventually I went exploring for its source. Unmistakably it was coming from the nesting hole. Perched on a ladder, I reached in tentatively. My arm went in all the way to my elbow, and my hand encountered something gooey. I pulled it out—a vile, stinking mass of putrefaction from dead hatchlings, broken eggs, maggots and rotting nest materials.

Not long after that we fled the city for our present island home. Here a small flock of starlings lives in the tiny "downtown" section of the island. We sometimes see them foraging in pastures and hayfields around the island, but we'd never had them in our woodland neighbourhood until four years ago. Then, one springtime morning, I heard that distinctive "quaint ditty"—a pair of starlings was setting about nest-building under the eaves of the house! I hurtled into action, spurred on by all the negative things I'd ever heard about this "black plague of the air," and by the vivid memory of that awful stinking nest I'd had to clean out. I didn't want to kill these newcomers, but I'd be damned if they'd get a toehold in our place. Wherever they set about nesting, I scrambled up and tore the nest materials away. They persisted in building. Equally determined, I persisted in tearing away. Eventually they gave up and disappeared, and the following spring did not return.

Then, this past spring, I was working at my desk when, gazing out the window, I saw they were back again: two chunky European starlings, waddling around on the lawn, pecking for grubs. I watched them. They're beautiful birds, really, their plumage a shimmering green-blue-black iridescence, the male's bill coloured bright yellow for courtship. When I went outdoors, the pair took off with that signature flight of theirs—the square tail spread and the short, pointed wings flapping rapidly to carry the stocky body. Minutes later they were back again, chattering and twittering with their creaky, guttural voices.

What to do? I looked to my prejudices. Starlings have a persistent reputation for displacing native birds. True or false? In *A Guide to Bird Behaviour* Donald Stokes says starlings, like blue jays and house wrens, are very aggressive in claiming nest sites. "In some cases other birds will be driven out of nest holes by starlings and their eggs and young eaten." Other books rendered the same verdict: these bold intruders regularly displace woodpeckers, martins, flickers and other cavity nesters.

Not at all aggressive the rest of the time, starlings get very territorial and very pushy at mating season. Typically, a male stakes out a nest site, a hole in a tree or building with a nearby perch. He defends his turf against other males by what Stokes calls "crowing"—tilting his head, fluffing out his throat feathers and emitting, "a continuous chortling call, variable and unmusical." Other intimidating manoeuvres include the fearsome "bill-wipe," in which the bird rapidly and repeatedly wipes its bill on the branch it's perched on, the dreaded "fluffing," in which all its feathers are puffed out in anger, and the menacing "wing-flick," where only the tips of the wings are extended and flicked rapidly. If a competitor dares to alight on his perch, the defender will place himself between the intruder and the nest hole and then begin "sidling," in which he sidles along the branch, gradually forcing the intruder off.

Yes, we could live with a half-dozen starlings in the yard, and would even enjoy having them around. But the grim lesson of history is that starlings, like humans, don't know when to stop. Adaptable and innovative, they are relentless in expanding their numbers and their range. Within sixty years of their introduction to North America, they'd increased a millionfold, and today they're found everywhere from Alaska to Mexico. Like ourselves, they tend to cluster in huge numbers in cities, towns and suburbs. After the young have fledged, the birds gather into enormous flocks for communal roosting in big shade trees, on buildings or under bridges. Roosts can sometimes contain a quarter-million birds. These two newcomers in our yard, I knew, weren't just a pair of gregarious characters looking for a quiet place to raise a family. No, they were the advance team, the shock troops, the thin edge of a wedge that could quickly widen into a catastrophic congregation of birds, capable of darkening the daylight sky and casting a noisy and dirty pall across our lives forever.

I pictured us in the same predicament as the mystified maintenance people at the Illinois State Capital Building in the fifties. They kept finding evidence of building stresses that they couldn't account for—until someone checked the roof and found it sagging under ten tonnes of starling droppings! I identified with bird-loving Britons whose starling flocks are said to sometimes number 100,000 birds.

I saw us going slowly mad, like some beleaguered farmers in B.C.'s Fraser Valley when wave upon wave of Vancouver's starlings alight each day to feed in the vegetable and berry fields. Soft fruits, particularly blueberries, as well as strawberries, raspberries, figs, cherries and sweet corn are favourite starling foods. A survey of farmers in forty-seven states concluded that starlings were the single worst cause of damage to grain crops. Livestock producers complained of starlings contaminating buildings, feedlots, farm equipment, grain elevators and cattle feeders. Other farmers reported starlings stealing large amounts of livestock and poultry feed. Holly growers, in Oregon and elsewhere, have had tremendous problems too, as there's a very limited market for holly foliage and berries streaked with starling droppings.

And as if this listing of offences isn't indictment enough, starlings have been convicted of causing at least one major plane crash—in 1960 a jetliner out of Logan International Airport in Boston collided with a flock of twenty thousand starlings and crashed to earth, killing sixty-two people. No wonder starlings rival rats for the dubious titles of "Public Nuisance Number One" and "This Country's Most Unwelcome Immigrant."

* * * * *

Beset by these unwelcome hordes, people have responded with everything from shotguns to plastic owls. As the Toronto shotgun corps discovered, there's a quality of infinity about shooting starlings; it's a bit like counting grains of sand on a beach. In *Pilgrim at Tinker Creek*, Annie Dillard retells an old tale of a countryman bothered by starlings roosting in a big sycamore near his house. He said he tried everything he could think of to drive them off, and finally in frustration blasted the roost with his shotgun, killing three starlings. "When asked if that discouraged the birds, he reflected a minute, leaned forward, and said confidentially, 'Those three it did.'"

For higher kill ratios, there are a number of avicides available, but poisoning the birds can create more problems than it solves. Some of the poisons that are very good at killing starlings are equally good at killing other birds. Also, some of them are known secondary poisons; that means that if a starling were to die and fall to the ground, and a dog or cat were to get hold of it, the pet could be poisoned as well.

Netting and trapping have been tried in different places. My pesticide handbook tells me that a single trap at an experimental station in Vineland, Ontario, "has taken many thousands of starlings each season for a number of years." What to do with a trapped starling? The handbook shows no mercy. "Disposing of the trapped birds may be accomplished by placing the trapped birds in a plastic bag with a few ventilation holes and connecting the bag by means of flexible hose to car or truck exhaust pipe. Remove the dead birds from the trapping site and bury."

If slaughter must be undertaken, I prefer the approach of the chef at the Harrison Hot Springs Hotel who entered the Centennial Culinary Arts Festival in Toronto with a dish called Pâté de Chasse—a pie filled with the meat of fifty starlings, along with veal, spices and brandy. I don't know how the entry fared, but back in Yorkshire the Reverend Morris was less than enthusiastic about starling cuisine: "They are good to eat, but rather tough and slightly bitter."

Recognizing the futility of trying to eradicate problem flocks, most people opt for a strategy of deterrence. Of course, there are the half-hearted repellent devices—fake plastic owls and smearing of Tacky-Toes and Roost-No-More pastes along roosting sites. These are entirely appropriate in certain circumstances: for example, a sticky paste was sprayed on trees to keep starlings from defecating along the route of Lyndon Johnson's 1965 inaugural parade. But let's face it: if twenty or thirty thousand starlings are roosting in a row of huge chestnut trees, a couple of tubes of repellent paste are a joke.

One solution—cutting the trees down—often doesn't go over well. In many cities, big shade trees grow on city-owned boulevards and faint-hearted politicians are loathe to provoke the wrath of tree-loving constituents. Instead, city planners look to a long-term solution of not planting tree species which provide roosts and replacing aging roost trees with more ornamental species. As an interim measure, heavy pruning of big trees can open up the canopy and reduce their attractiveness to starlings. Similarly, new buildings and bridges are now designed so that they don't provide roosting ledges for starlings, pigeons and sparrows. Home-owners are urged to check their buildings for potential nesting sites under eaves or in nooks and crannies. Galvanized wire can be used to close off access to such areas.

These are long-term solutions, but short-term deterrence is much more fun. In its full glory, it works upon the principle of creating such a bedlam of sound and light that the birds simply move away to get some peace and quiet. I think it's derived from the old military strategy of liberating a village by blowing it up. A wild spectacle of sight and sound is what's called for: scarecrows, streamers, reflectors, balloons, strobe lights, firecrackers, noisy gas guns, vibrators, ultrasonics, recorded distress and predator calls—almost anything you can think of. Starlings have one of the widest hearing ranges of any bird, and noise is the primary weapon in the starling chaser's arsenal. Back in Illinois, an enterprising radio station came up with the bright idea of broadcasting a tape of a dying starling and urging all listeners to place their radios outdoors, tune in and turn up the volume! A number of cities have employed a boom box to play amplified starling distress calls on tape which, in the graphic phrase of one reporter, sound "like a large pig being slowly skinned alive."

But no single technique, however clever, works for very long; to be effective with starlings you need a constant barrage of noisy new tricks, like late-night TV advertisers. In Vancouver, the city's starling squad strings helium balloons above the roost trees and flashing lights through the tree branches. Flare guns are fired. The skinned pig distress call screeches. Strobe lights flash. A propane cannon periodically erupts with a huge BOOM! After several nights of this madness, the starlings abandon the neighbourhood.

* * * * *

But amid this tumult, is there nothing good to be said for starlings? Their worst offense, like our own, it seems to me, is what one naturalist called their "unbounded fecundity." They are extremely good parents, "very assid-uous in their care of their young," as Morris puts it. The male establishes a nest, cleaning out the hole and lining it with bits of leaves, bark, mosses and lichens. Stokes says the male will often perch outside his nest with a piece of nest material in his bill in order to attract a mate. He'll drop and retrieve the scrap, and later carry it in and out of the nest, all as an elaborate demonstration of his worthiness as a provider. Once a female chooses a nest and the male that goes with it, she'll throw out much of the nest he's built and redo the job properly while he perches nearby.

After that it's all togetherness, as the birds share their daily courtship activities, feeding, roosting, perching and flying as a pair. The clutch of four or five eggs—much like robin eggs in size, shape and colour—are incubated for twelve days by both parents, though only the female sits on them at night. After hatching, the ravenous nestlings are fed by both parents for about twenty-three days before fledging. Soon after fledging, the mousey-brown young birds join with other juveniles to form huge flocks that move together about the countryside feeding in open fields. Often the parents will get right back to work on a second brood, beginning around midsummer. After this second batch has fledged, parents, juveniles and unmated birds all unite in massive late-summer flocks. In autumn the birds undergo a complete moult, taking on a temporary speckled look from the white-tipped new feathers. By then the big flocks are already breaking up and perhaps changing locations. Some adults already start next year's breeding preparations during winter.

A second mitigating factor in starlings' favour is that, while they eat and contaminate a lot of valuable foodstuffs, they also consume enormous num-bers of pest insects. A group of Russian scientists determined that about 75 per cent of the nestlings' diet consists of harmful insects. Another study of two thousand individual starlings found that over half their food was insects such as beetles, weevils, grasshoppers, wood lice, millipedes and caterpillars. Waddling about on lawns and gardens and pecking assiduously, they glean tremendous numbers of grubs from the ground, particularly the troublesome grubs of Japanese beetles.

A characteristic that starlings—close relatives of the mynahs—share with ourselves is that they are great communicators. Stokes calls their song "an extremely variable but somewhat melodious set of calls, often incorporating shrill squeals, squawks, and many imitations of other birds' calls and songs." Reverend Morris liked it too: "On sunny days they may be heard gurgling a low and not unpleasant note, which, when the result of the 'concerted music' of a flock, forms a body of sound to which you like to listen."

Well, maybe. We were saved this past spring from having to make any more hard choices about whether to prevent starlings from nesting in our yard: the pair disappeared voluntarily. Perhaps the great horned owls in our woods scared them off, the way the plastic owls are supposed to. More likely they just found better accommodations elsewhere. Ultimately, there may be something to be said for having them about, presenting an iridescent mirror that reflects back the unpleasant consequences of an aggressive, noisy and polluting species whose population is out of control.

Chapter Fifteen

SNAKES

Saint Patrick's Malediction

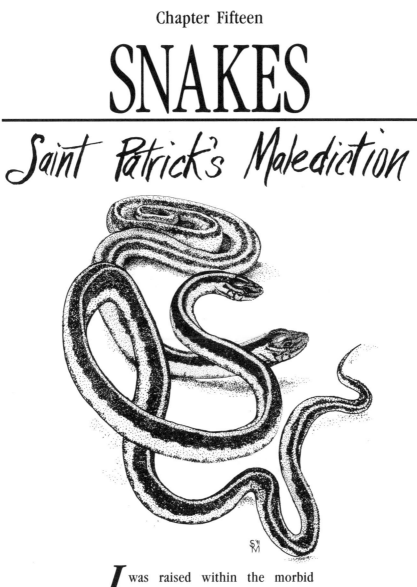

I was raised within the morbid confines of strict Irish Catholicism. Black-cassocked priests thundered from their pulpits about mortifying the flesh in order to resist the temptations of the Evil One. I tried my best. Often we pondered the seduction of Eve by the serpent, and her seduction of Adam, causing humankind's fall from God's grace and banishment from Eden. Saint Patrick was our patron, the holy man who collected all the snakes in Ireland, placed them in a box and hurled the box into the Irish Sea, so that when the sea ran rough the old people said it was Saint Patrick's snakes writhing to escape their box!

Sitting stock-still in church every Sunday morning, or creeping into a confessional on the Saturday before in order to whisper my sins of the flesh into the dark, I had no idea there was any other side to these serpentine stories. The nuns never told us how, in the fifth century, Saint Patrick was locked in combat with Celtic druids, nor how an Old Irish hymn to the Saint sings of protection against "women, smiths and druids"—the witches, alchemists and nature-worshippers possessed of magic powers.

There was no mention in catechism class of the snake ring that holds its tail in its mouth, a magic creature which the alchemists called Uroboros; and many years passed before I read in Nor Hall's *The Moon and the Virgin* an entirely different version of the snake.

> Because of the open feminine mouth and the phallic tip of the tail and the miraculous way it sloughs its skin in yearly regeneration, the uroboric snake is an enchanting creature containing all possibilities of male and female, beginning and end, life and death.

Living now in a garden crawling with snakes, on a different western isle where "women, smiths and druids" go unmolested, I've come to suspect (forgive me, dear Sister Rosalie!) that snakes got caught up, not in some devil's bargain, but in a nature-hating and women-hating ideology that has left us sadly estranged from our earthly home.

This is a heavy theological load, I know, to place upon the non-existent shoulders of the snake. Fear and distrust of these cold-blooded reptiles seems entirely reasonable; coming upon one suddenly, we feel what Emily Dickinson so perfectly described as "a tighter breathing/and zero at the bone." I wouldn't argue that our fears of this "narrow fellow in the grass" are unfounded; it's estimated that perhaps as many as forty thousand people die from snakebite every year, mostly in Asia, Africa and Latin America. Venomous snakes come in all sizes, up to the four-metre deadly king cobra. There are authenticated records of young humans being swallowed whole by huge pythons in Africa and Asia. The longest snake ever reliably recorded was a monstrous reticulated python shot in the Malay Archipelago in 1912—it measured more than ten metres. With all sorts of venomous snakes on the loose, and with huge pythons, boas and anacondas reaching ten metres and weighing two hundred kilograms, fear of snakes, it can be argued, isn't a consequence of misguided theology, it's a matter of plain common sense!

But is it? Long before Saint Patrick's time, long before Christianity, the snake was a powerful and dangerous presence in the religious observances of many tribal peoples in Africa, Asia and the Americas—but remnants of the snake cults indicate a fear tempered with respect, and ritual behaviour intended not to destroy the snake, but to live in harmony with it. Funk & Wagnall's *Standard Dictionary of Folklore, Mythology and Legend* describes a spring snake ritual of the Yakut Indians of central California.

> Shamans, or men with special powers to whom rattlesnakes had spoken in their dreams, lured the snakes out of their dens in early spring, took them home and at a public ceremony placed the snakes, in a sack, on the heads of spectators. The next day the shamans "cured" these spectators of prospective snake bites by the usual sucking method. After taking these precautions for the benefit of the community, the snake shamans played with the snakes, threw them about, allowed them to bite their thumbs or hands, and so forth.

Afterwards, the shamans placed the snakes in a small hole outdoors and the people prodded them with sticks. The shamans paid the people shell money not to harm the snakes. Then each member of the community filed past the hole and placed the right foot into or over it. This "stepping" rite ensured that for a year every snake, when approached, would rattle in warning instead of striking.

Another spectacular performance with snakes, noted in the same source, is the Snake Dance of the Hopi pueblos in which live snakes, grasped by their middle in the dancer's mouths, are carried around the plaza.

> The snake dance, like other Hopi ceremonies, is celebrated largely as a prayer for rain. The origin myth attached to it recounts how Snake Hero and Snake Maid had children who were transformed into snakes; hence the Hopi regard snakes as their elder brothers, and powerful in compelling the Cloud People to bring rain.

Similar beliefs and rituals occurred elsewhere. In *The Golden Bough*, Frazer records how certain Australian aboriginals also employed snakes in rain-making rituals, and how a snake clan of ancient Africa exposed their infants to snakes "in the belief that the snakes would not harm true-born children of the clan."

Elsewhere in Africa, young women were ritually married to a snake god, the marriages being consummated by medicine men, and any resulting children being revered as the children of God. In Mexico, Africa and Asia, snake dances with long, serpentine lines of dancers invoked the serpent's magical powers of fertility for earth and rain. "If you obey the snake, you and yours shall thrive," sang members of the snake tribes of the Punjab during ritual processions. "The snake tribe is not uncommon in the Punjab," Frazer goes on. "Members of it will not kill a snake, and they say that its bite does not harm them. If they find a dead snake, they put clothes on it and give it a regular funeral."

Similarly, Frazer continues, many American natives would not molest snakes when they came upon them, believing that if they were to kill one, other snakes would kill some of their relatives or friends in retaliation.

So the Seminole Indians spared the rattlesnake, because they feared that the soul of the dead rattlesnake would incite its kindred to take vengeance. The Cherokee regard the rattlesnake as the chief of the snake tribe and fear and respect him accordingly. Few Cherokee will venture to kill a rattlesnake, unless they cannot help it, and even then they must atone for the crime by craving pardon of the snake's ghost either in his own person or through the mediation of a priest.

If the ritual atonement was neglected, it was thought the kinfolk of the dead snake would avenge its death by sending an emissary to seek out and kill the murderer.

With the snake representing magical powers over life and death, as well as fertility rites whose pathetic remnants we now see in exotic dancers fondling pet snakes while taking off their panties in front of drunken men, it's no wonder good Saint Patrick and the others decided to declare war upon the serpent!

* * * * *

Where I live the only dangerous snakes are mythological beasts—the double-headed snake *Sisutl* of the Kwagulth people, and *Haietlik*, the lightning snake of the Nuu-chah-nulth, supposedly used as a harpoon by the Thunderbird when hunting whales. For the rest, we have to make do with several species of garter snake. On hot summer days we see dozens of them in our flower and vegetable gardens, sunning themselves on the top of rock walls, poised motionless at the edge of a little pool, or slipping silently through the grass. Striped and splotched with reds, blues, greens and yellows, I can't tell one species from another, and herpetologists often have to closely examine the scales and plates on a garter's head to accurately determine its species.

The most widespread snake in all of North America is the common garter snake (*Thamnophis sirtalis*), which ranges from the Pacific to the Atlantic and from the Northwest Territories to the Gulf of Mexico, though not through the American southwest. Long and slender, the common garter can reach 120 centimetres, and with its flat head, rounded snout and large eyes, a big specimen can look fairly formidable. Sometimes called the water snake because they're often spotted along creeks and lakes or near the sea, common garters feed largely on water creatures—salamanders, frogs and tadpoles, small fish and fish eggs. They'll readily take to water, and we often see them swimming smoothly and rapidly across small lakes and ponds.

Most of the specimens in our gardens, I think, are northwestern garter snakes (*Thamnophis ordinoides*). This is a smaller species, growing up to about forty-five centimetres, which lives west of the coastal mountains from B.C. south to California. At home among dense thickets along the edges of roadways, fields and other clearings, it's usually the most abundant snake in its range. Scientists believe that the last ice age drove all reptile species southwards and that, when the glaciers retreated, the reptiles recolonized northwards. Today garter snakes are the only reptiles which live north of the fifty-first parallel, and population densities increase the farther south they are.

* * * * *

Every so often I come across a snake with a large banana slug in its mouth—a slug far fatter than the snake itself. Sometimes you'll see one comically caught with a slug half-way down its gullet, its whole head bulging with the effort. Or we'll see a snake basking in the sunshine with a conspicuous lump in its midriff, evidence of a large meal recently consumed. A snake can in fact swallow prey of much greater diameter than its own head or neck, thanks to a swallowing apparatus in which jaws, ribs and skin are so hinged and connected that they articulate with a contortionist's elasticity. An old handbook from the Royal Ontario Museum describes how a snake stuffs such large prey in.

Swallowing is accomplished by a "walking" movement of the jaws, those of one side being first pushed forward over the prey and then drawn backward while those of the other side are pushed forward. The needle-shaped, backward projecting teeth of each side naturally release their hold and slide over the prey during the forward movement and grip during the backward pull. In this manner the snake literally pulls itself outside of its victim which is forced down the gullet by muscular contractions and slight sidewise bendings of the body.

Biologists point out that this swallowing mechanism is just one of many highly specialized structural adaptations snakes have made. We tend to think of them as simple and primitive reptiles, unchanged from the wily serpent who seduced poor gullible Eve, but science says it isn't so. "For the past 20 million years," writes Harvey Lillywhite, a professor of zoology at the University of Florida, "snakes have undergone an impressive adaptation radiation, attaining a wide diversity in terms of habitat and behaviour." At home in everything from tree tops to oceans, the twenty-seven hundred species of snake have, says Lillywhite, learned to live in virtually every available habitat except polar regions, high mountaintops and deep oceans.

One question that particularly intrigued him, but might not occur to the casual observer, was how tree-climbing snakes maintain their blood circulation when their long, thin bodies are extended vertically, as when going up a tree trunk. What prevents blood from pooling at the tail, causing blood-loss to the brain? What he and his colleagues discovered, he writes in *Natural History* magazine, was that snakes have evolved different heart and lung arrangements according to the gravitational environment in which they live. In sea snakes—some of which may dive as deep as ninety metres—the heart is located mid-body. But in semi-aquatic species it's farther forward and in terrestrial and arboreal species it's farther forward still. "The forward heart," writes Lillywhite, "reduces the distance to the brain, so that when the snake is in an upright position, the weight of the blood column above the heart is less of a challenge." Besides this anatomical adaptation, he continues, the flexible body of the snake allows it to form itself into loops and drape itself over tree limbs in order to combat the stress of gravity and control blood circulation. In this respect, says Lillywhite, limbless snakes have more control over blood circulation than do such habitually erect animals as the giraffe.

These conclusions have badly undermined that stock snake scene in every

B-grade jungle movie ever made—the one where a huge constrictor slips from an overhead branch to coil about some unwary character passing below. In reality, arboreal constrictors are slender snakes averaging little more than two metres in length. "Contrary to popular belief," writes Lillywhite, "the giant serpents, which reach lengths from fifteen to nearly thirty-three feet, are mainly terrestrial or semi-aquatic, and none are primarily arboreal. The image of giant pythons dropping onto people from trees is simply a myth created by Hollywood moviemakers."

* * * * *

Death from above is more likely to strike the snake itself. I was out repairing a broken water line in the yard one spring afternoon when I heard a sudden rustle close by. I looked quickly to see, about ten metres away, a great horned owl taking off with a garter snake clutched in its talons. I've seen bald eagles do the same thing on the beach. Birds are the garter's worst enemy and in defence it must rely upon concealment and escape.

Northwestern garter snakes are at a distinct disadvantage here, because they live in a climate which requires that they regularly expose themselves to predatory birds. The climate is generally cool and moist, and much of this snake's food supply—slugs, earthworms and salamanders—is most readily found on damp days. "Whenever a snake eats a meal, begins to shed, or is pregnant, it must bask to absorb what little sunlight filters through the clouds and mist," writes Oregon herpetologist Edmund D. Brodie III. Brodie spent four years studying the escape behaviour and colour patterning of northwestern garter snakes, and summarized his findings in *Natural History* magazine in 1990. A snake's need to bask in order to maintain body temperature can, in a cool, moist climate, add up to a large amount of time exposed to potential predators. Since many hunting birds look for colour and movement, a snake's colour pattern could, writes Brodie, "determine whether it lives or dies."

Herpetologists have concluded that snakes with striped patterns usually live in open areas such as meadows, are active during the day, forage widely for their food, and when threatened try to escape rather than fight. The best most garter snakes can do when molested is secrete a foul-smelling blob from scent glands in their tails. By contrast, Brodie continues, blotched or spotted snakes, like rattlers, copperheads and cottonmouths, tend to be generally secretive, hunt at night, and become aggressive when threatened. The colour pattern and threat response appear to be correlated, each creating a different and effective optical illusion. Stripes on a moving body give the predator no point to fix on and make it more difficult to detect motion or judge speed. Spots, on the other hand, are more concealing on a stationary body, blurring its outline. Striped snakes do better if they move steadily away, blotched ones if they freeze and move intermittently.

What fascinated Brodie is the exceptional variety of colours and markings, including both stripes and blotches, found among northwestern garter snakes, even from the same litter. Why such variety, and how does it affect escape behaviour? After examining more than twelve hundred offspring of two hundred garter families, he concluded that a northwestern garter's colour pattern and escape behaviour come as a package. "Striped snakes usually crawl straight away from a predator, while spotted or unmarked snakes often reverse direction and freeze." How does a snake know which tactic to employ? Appearance and behaviour, concludes Brodie, must be genetically linked— the genetic coding for stripes must be inherited along with an instinct for direct escape. When, near the end of summer, a half-dozen or so small garters are born live and squirm free from their mother, each sheds its membranous sac and emerges with a distinctive colour pattern which will remain unchanged for life and which, with its corresponding behaviour, will determine how long that life may be.

* * * * *

The beauty of snakeskins has not been lost on fashion-conscious fops. The *Endangered Species Handbook* notes that "lizard and snakeskin products are now taking the place of turtle and crocodilian leather in the luxury trade. Handbags, wallets and shoes from these reptiles can be seen in department and shoe stores in this country as well as Europe and parts of Asia." Millions of snakes are being killed worldwide to supply the market, and in one year Indian Customs officials confiscated 150,000 snakeskins being illegally imported. The handbook says the boa constrictor and all pythons have been severely depleted and are now listed in the Convention on International Trade in Endangered Species.

How much wiser and more civilized was this approach mentioned in *The Golden Bough*:

> The Huichol Indians admire the beautiful markings on the backs of serpents. Hence when a Huichol woman is about to weave or embroider,

her husband catches a large serpent and holds it in a cleft stick, while the woman strokes the reptile with one hand down the length of its entire back; then she passes the same hand over her forehead and eyes, that she may be able to work as beautiful patterns in the web as the markings on the back of the serpent.

That's a kind of reverence for the natural world we somehow lost, and whether or not the losing was the fault of Saint Patrick and his ilk now seems less important than re-establishing the connection. Snakes, I think, are a good starting-point for rediscovering Eden. In our gardens at home we've taken particular care to maintain suitable snake habitat. In clearing our homesite we left in place a number of the huge old cedar stumps beneath which snakes seem to find protective crannies. The many dry stone walls we've built provide great snake habitat, and in garden corners we heap up waste stones dug from the flower and vegetable beds to provide shelter and hibernation places. We use no poisons for slugs or anything else. As a result, our snake population keeps growing—you can't walk down a garden pathway without seeing several of them, forewarned of your approach by vibrations on the earth, slipping surreptitiously under nearby foliage. There is a surreal and thrilling beauty in their elaborate markings, the near-sighted staring of their lidless and always-open eyes, and the flickering of their delicate forked tongues licking the air for telltale odour particles. It's not hard to understand how we came to hate and fear them, or to see that our hatred and fear of nature must at last be put to rest.

Chapter Sixteen

STINGING NETTLES

Devilishly Delicious

W e were listening one day to a
favourite gardening program on
the radio, one in which gardeners and
would-be gardeners call in with problems and questions for a gardening
expert. A caller came on the line to say
that he'd just bought a lot on one of
the Gulf Islands but the property was

infested with stinging nettles. How could he get rid of them? The expert prescribed a herbicide, then, as an afterthought, mentioned that "some people eat nettles in the spring." But by then the caller had gone, presumably speeding to a garden shop to buy a bottle of weed-killer.

I know what this chap was feeling, and he could have been me twenty years ago, because when we first moved onto our island acreage it too had an infestation of stinging nettles. Among our earliest experiences of "getting back to the land" was brushing carelessly past a patch of nettles and flinching in alarm at their fiery stings. Within minutes a welter of little white blisters bloomed on our reddened skins and a powerful itching set in. We consulted our homesteading bibles; one advised that wherever stinging nettles grow, a dock plant grows nearby, and the itch of stings can be relieved by chewing a dock leaf and applying it as a poultice to the inflamed area. This old remedy apparently underlies the jingle, traced back to pre-Elizabethan England:

> Out nettle, in dock,
> Dock shall have a new smock,
> But nettle shan't have none.

Smocks or not, we'd had our first brush with nettles, and learned to both despise and avoid them. They seemed to be growing everywhere on our place—in shady woodland glades and openings, in damp hollows and along the little creek, and edging the margins of the forest. By summertime their clustered stalks would reach two metres high, giving the place the look of a grim wasteland. As we started clearing ground for our first gardens, we found nettle roots, long thin rhizomes spreading like translucent yellow webbed nets just beneath the earth's surface.

On each tall nettle stalk, opposite pairs of leaves grow on short stems, the pairs alternating in their direction of growth. Heavily veined on the underside, the leaves are roundish at their base with jagged-toothed edges tapering to a sharp point. Up to fifteen centimetres long and vivid green in spring, the leaves pale and become ragged and tattered as the season wears on. In early summer, clusters of small flowers, greenish and nondescript, hang like tiny braids from where the leaf-stems join the stalk. Upon careful inspection you see that each leaf edge, stem and stalk bristles with fine hairs. Hollow, like tiny hypodermic needles, each hair is filled with a fluid related to oxalic acid and tipped with a microscopic and brittle bulb. When touched, the bulb shatters and its jagged edges scratch your flesh, depositing the irritating fluid. *Urtica dioica* is stinging nettle's proper name, the generic name *Urtica* meaning "to burn or sting"; in Ireland it was said that the stings were pricks from the Devil's pitchfork!

It only took a couple of close encounters with these noxious plants to put me in the same frame of mind as the pesticide-questing radio caller. Eradication seemed the order of the day. Nettles are listed in government handbooks as "poisonous and injurious weeds" capable of causing skin irritation. A British weed-control handbook lists them among the country's "most troublesome weeds," accusing them of "presenting a hazard to road safety and/or to agricultural crops." The road hazard, I assume, comes from nettle's habit of growing profusely along the verges of country lanes and obscuring driver vision by leaning en masse over the carriageway. The British book recommends spraying with 2,4-D as the best way to control this pest plant.

We were saved from that sort of folly by a visit from my parents, both of them steeped in rural Irish working-class lore. I think on their first visit they were more excited about our tattered patches of stinging nettles than about the magnificent stands of cedar and fir trees on our place. They talked about how to cook nettles in the spring and, after returning to Toronto, my mother sent out her old recipe for nettle beer. We began to realize that nettles are more than just the "troublesome weed" of agricultural handbooks. And who should pop up again but good Saint Patrick himself! As Lesley Gordon writes in *A Country Herbal*, "St. Patrick blessed this unloved plant for its services to man and beast, for the nettle has been put to more varied and useful purposes than almost any other herb."

Four hundred years ago, a British writer named Coghan, in *The Haven of Health*, passed on to his readers the same advice my parents gave.

> I will speak somewhat of the nature of Nettle that Gardeners may understand what wrong they do in plucking it for weede, seeing it so profitable to many purposes. . . . Cunning cookes at the spring of the yeare, when Nettles first bud forth, can make good pottage with them, especially with red nettles.

* * * * *

Virtually penniless in our early homesteading days, and determined to "stalk the wild asparagus" and any other wild food gratis for the gathering, we leapt upon this opportunity for a free springtime feast. Early settlers called it "Indian spinach." Nowadays, one of the very first food-gathering rituals of our year is getting a colander of new nettle-tops for steamed greens. If there's a warm spell in February, as often happens, the earliest nettles emerge in sunny, sheltered spots. Within days they're eight centimetres tall and ready for picking. At that point their tops are tight little rosettes of purplish green leaves. On the first few days of snipping them with scissors, I always have a sense of exhilaration at being in touch with the earth again, with the sweet scent of warming soil, the thrill of living things awakening after the entombment of winter. No matter that there'll be more soaking southeasters to come, more frost and snow and dreariness. In nettles, before almost anything else, the forces of spring have awakened, the juices of life have begun to surge, and intimations of later abundance are given.

In his *Complete Herbal*, Nicholas Culpeper explained why nettles are so good to eat in the spring.

> This is also an herb Mars claims dominion over. You know Mars is hot and dry, and you know well that Winter is cold and moist; then you may know as well the reason why Nettle-tops eaten in the Spring consume the phlegmatic superfluities in the body of man, that the coldness and moistness of Winter hath left behind.

Whether that's exactly the reason, I'm not sure, but it's certain that a steaming helping of springtime Indian spinach exudes healthfulness and vigour. Gerard's *Herbal* says that eating nettles baked in sugar "makes the vital spirits more fresh and lively," and that's just how I feel. Most often we just pour a colanderful of tops into a pot of boiling water. Once it has boiled over the leaves, we'll drain the water off, saving it to cook rice in or to drink as a cold tea. The nettles are so tasty on their own they don't need much seasoning—a pinch of salt and lemon pepper, a touch of garlic and a squeeze of lemon juice. Mozzarella cheese melted over them is an optional extravagance. Baked in casseroles or as a substitute in spinach pie, they combine wonderfully with wild chanterelle mushrooms which we've picked and processed the previous fall. Nettle soup is another springtime standby. We simply make a creamy soup base with butter, flour and milk, and then add nettle tops that have been steamed, drained and cut up. Simmered, the soup is seasoned with herbs, spices and miso. Some recipes add chicken stock, egg yolks and heavy cream.

"We did eat some nettle porridge," wrote diarist Samuel Pepys in 1661, "which was made on purpose today for some of their coming, and was very good." Lesley Gordon writes that nettle porridge was a common dish in seventeenth-century England. In fact, in those days an accomplished cook was expected to have at least seven good recipes for using nettle-tops. In

Scotland, says Gordon, nettle soup and nettle pudding were common dishes. The pudding was made by mixing nettle-tops with rice and a brassica— cabbage, broccoli or Brussels sprouts—cooking the mix in a muslin bag and serving it with butter.

Nettle tea and beer have also been popular with countryfolk for centuries. The tea is brewed simply, by pouring a litre or so of boiling water over about five handfuls of young tips and allowing them to steep for several hours. My mum's old recipe for nettle beer has disappeared somewhere over the years, but it was very similar to the one Lesley Gordon gives.

Nettle beer was a pleasant country drink made of nettle-tops, dandelions, goosegrass and ginger, boiled and strained. Brown sugar was added, and while still warm a slice of toasted bread, spread with yeast, was placed on top, and the whole kept warm for six or seven hours. Finally, the scum was removed, a teaspoon of cream of tartar was added, and the beer was bottled.

This concoction was consumed to ease the pain of gout and rheumatism, as well as for refreshment. Nettle tea, too, has a distinguished medicinal history. Many cultures used it as a spring tonic and blood purifier. Funk and Wagnall's *Standard Dictionary of Folklore, Mythology and Legend* reports that "the Chippewa Indians of the U.S. use a mixture which they call Winnebago medicine made of stinging nettle root and lady fern root for disorders of the urinary system." In old Anglo-Saxon medicine, says the dictionary, "the nettle was considered effective against the green venom, one of the nine flying venoms causing disease." The great preacher John Wesley recommended this remedy: "Boil nettles till soft. Foment them with the liquor, then apply the herb as a poultice. I have known this cure a Sciatica of forty-five years standing."

Ancient Romans, American Indians and others used nettles as treatment for the common cold. It was also taken as a cure for haemorrhoids. Culpeper proclaimed nettles effective against wheezing, shortness of breath and sore throats. He recommended a tea made from the seeds, which would "killeth worms in children, easeth pains in the sides, and dissolveth the windiness in the spleen, as also the body, although others think that it is only powerful to provoke venery."

This last was a reference to nettle seeds' reputation as an aphrodisiac, supposedly a firm belief of some German peoples. But they also believed that nettles gathered before dawn were good for ailing cattle. Toothache was another ailment treated with nettles. A compilation of Aztec herbal practices entitled

The Bodianus Manuscript of 1552 recorded that "pain in the teeth and gums is allayed by scarifying the gums and cleaning them of pus, if to the festering part be applied the seed and root of nettles, ground in the yolk of an egg and a little white honey." In the colonial days of what was to become British Columbia, the pioneering doctor William Tolmie found himself unable to cure an outbreak of scurvy in 1835, until he visited a native encampment and "brought home a quantity of Nettles and the herbaceous plant so much eaten by the natives."

* * * * *

Besides these brilliant medicinal and culinary careers, nettles have been employed for everything from conditioning the hair of Victorians to exorcising the Devil. Several strains of European folklore identified both thistles and nettles as the Devil's vegetables, and in Yorkshire nettle leaves were used in an exorcism ritual. In parts of England and elsewhere a common superstition held that nettles kept lightning from striking nearby. At one time the plant was considered sacred to Thor, God of Thunder, and was worn as a sort of amulet against the fear of danger. And nettle could banish cold as well as fear—Lesley Gordon quotes an ancient prescription which she thinks may date from the Roman occupation of Britain: "Take nettles, and seethe them in oil, smear and rub all thy body therewith; the cold will depart away." There's an old tradition that Roman soldiers stationed in this dank and thankless outpost of the Empire brought from sunny Italy the seeds of Roman nettle (*U. pilulifera*) and used the leaves of the plant to chafe their bodies to stimulate blood circulation and thus stay warm.

In the nineteenth century a nettle hair conditioner was concocted by simmering young tops for several hours, straining and bottling the resultant tea. I haven't seen this yet on the "natural products" shelves, but it can't be far away. Beatrice Trum Hunter writes about another use: "Fruit packed in nettle hay is hastened in its ripening. This weed deters fermentation, keeps fruit free of mold, and bestows upon it good keeping qualities." As early as the seventeenth century, orchardists were using the plants as packing for plums, apricots, peaches and other stone fruits.

But by far the most important and widespread use of nettles, other than as food and medicine, was as a fibre for textiles. In one of his stories, Hans Christian Andersen shows a little princess spinning nettle fibres, and in fact the spinning of nettle-flax for linen was common throughout Europe before imported cotton became available. "I have slept in Nettle sheets," wrote the nineteenth-century poet Thomas Campbell, "and I have dined off a Nettle table-cloth ... the stalks of the old Nettle are as good as flax for making cloth. I have heard my mother say, that she thought Nettle cloth more durable than any other linen." Some American Indians made nettle rope from the tough, stringy fibres, and Gordon writes that in Europe they were also traditionally used to fashion rope, thread, sailcloth, sacking, paper and twine for fishing nets.

* * * * *

For about a dozen years we kept a small herd of dairy goats on our place, and one of the joyful Thomas Hardy-type rituals of summer was scything down great swaths of nettle and letting them dry in the sun like hay, then bundling them by hand and storing them in the barn for winter feed. They were one of the goats' favourite delicacies (almost as good as our rose bushes!), and I'm convinced they helped keep our vet bills to a minimum. Often when we're out gathering nettle shoots in spring, we'll find dozens of banana slugs grazing on the young plants too. A number of butterfly species are dependent upon nettles for the early stages of their life cycle—in Europe, nettles are the only food eaten by the caterpillars of tortoiseshell and peacock butterflies—and widespread eradication of nettles is said to be threatening some butterfly populations.

Similarly, many writers have praised nettles for their beneficial effects upon other plants. "Stinging nettles seem to have a number of helpful qualities, changing the properties of neighbouring plants and making them more insect-resistant," writes Beatrice Trum Hunter. "The iron content of nettle helps plants withstand lice, slugs, and snails during wet weather. Mint and tomatoes are strengthened in the vicinity of stinging nettle ..." Another true believer, Lesley Gordon writes that "today it has been proven that stinging nettles help neighbouring plants grow more resistant to disease. As a companion plant it increases the content of essential oils in neighbouring herbs, and stimulates humus formation." As well, nettles have a reputation for activating compost heaps, and whenever possible we include a few layers of fresh-cut nettles when making our compost piles. Similarly, we'll put a layer of them in the compost privy every now and then to stimulate decomposition. A complete liquid fertilizer can be made by soaking nettles in a barrel of water for a couple of weeks. Audrey Wynne Hatfield writes in *How To*

Enjoy Your Weeds that this nettle tea "is not only a good folia feed, but also an effective spray against mildew, black fly, aphis, and plant lice."

* * * * *

Nowadays, I worry not about how to get rid of stinging nettles, but how to keep an abundant supply on hand. They don't really like being cut back too frequently or vigorously and, at least in places, are prone to being pushed out by other plants. On our property they seem content to congregate along fencelines and around compost heaps and places where they can spread their roots under boards or other surfaces. They're also quick to leap into areas where the earth's been disturbed the year before. Convinced of their beneficial effects, we encourage nettles to surround our vegetable patch. They're easily removed by hand, shallow roots and all, from any place we don't want them. As a staple of our diet from February through May—an early and tremendously healthful plant that requires no seeding, watering, cultivating or other care—and as a dried tea thereafter, it would seem entirely foolish to eradicate this valuable crop and then fuss over vegetables that require much more care and offer far less by way of vitamins and minerals.

Nettle stings are easily avoided, although city visitors often need to be forewarned, and even kids around our place quickly learn to co-exist with nettles. Occasional stings can be treated, if not with the traditional chewed dock leaf, then with calamine lotion, moistened baking soda or just plain mud. Urine and salt water were traditional native treatments. I used to wear gloves when gathering the young tips in spring, but now I do it barehanded. By grasping the tips firmly, I'm able to handle dozens of plants and suffer nothing worse than a pleasant tingling in my fingertips.

By late summer our nettles are a tattered-looking bunch, their leaves chewed by slugs and caterpillars, their seed clusters brown and burnt, their long stalks bent over by wind and rain. They're wild and rangy and a little bit scruffy, not at all the sort of thing one would want in a primly manicured garden. But they're a wonderfully useful and beneficial plant, and it's sad to see so many people scrambling to get rid of them without a moment's thought. I'm glad we have them around our place, not just for their practical uses, but because they're part of a complex plant and animal ecology that we can't even begin to understand. Ultimately, nettle is another of those species that cry out to us to re-examine our mania for absolute control of our landscapes, our need to eradicate certain fibres in the web of life, when their full role in that web is something of which we are only very dimly aware.

Chapter Seventeen

RATS

Who's Racing Whom?

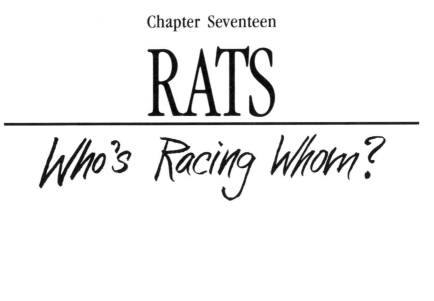

"*E*ating rats can open up a new supply of meat, turning a harm into a benefit." This was the upbeat assessment of China's official *Economic Information Daily* in early 1991. The paper was trying to stimulate interest in rodent meat, available in Chinese markets for twenty-five yuan ($5.75) a kilogram. Besides the traditional preparation by roasting, stewing, smoking, stir- or deep-frying, the paper was urging its readers to widen the horizons of their rodent cuisine with dishes such as rat steamed with lotus leaves or with chestnuts and bamboo shoots. Or fried with asparagus. How about a hearty bowl of rat soup?

Some would see all this as a delicious bit of table-turning on the rats, which have been freeloading off human food supplies for far too long. And who better than the Chinese to initiate such a campaign? For it was China which likely gave the world its worst and most prolific rat. The "Year of the Rat," 1984 got a lot of coverage in the Chinese calendar, but in the previous year Beijing hosted one of history's biggest rat kills. With rodents running out of control in the city, officials launched a counter-attack which saw over 2 million rats dispatched. Residents hunted them ruthlessly, killing over 76,000 per day. One neighbourhood group was credited with, "wiping out more than 7,000 rats by baiting and beating." Reports said soldiers clubbed about 400,000 rats to death, while construction workers clobbered them in the depths of the city's labyrinthine sewer system. By the time the slaughter had stopped, the rat population was supposed to have been cut by as much as 95 per cent in some districts.

The rat has been called the most successful biological opponent of humans on earth. We've trapped it, poisoned it, bred dogs and kept snakes to kill it, studied it obsessively, done everything we can think of to defeat it—and failed! There are about 4 billion rats on earth—roughly one of them for every one of us—according to the World Health Organization. Very much our mirror image, they're smart, organized into strictly disciplined social units, clannish, tremendously adaptable, ferocious in combat. In short, like ourselves, they're survivors. It's not without reason that we call the daily grind of our lives "the rat race."

We hate them as competitors and speak of them with disgust. When we smell a rat, we sense there's treachery afoot. To rat on someone is to squeal, to become the lowest sort of informer. Strikebreakers used to be called rats the way they're now called scabs. "You dirty rat!" we snarl at adversaries. Inextricable entanglements become a rat's nest. Desperation forces one to fight like a cornered rat, while hopeless dejection makes one look like a drowned rat. Final failure and defeat leads you to exclaim, like Charlie Brown, a simple and pathetic "Rats!"

* * * * *

But what rats are we talking about? There are hundreds of species worldwide—pack rats, rabbit, rice and rock rats. There are water rats and African swamp rats and a giant Sumatran bamboo rat. We lump them all together, equally odious, equally beneath contempt—without just cause, of course. Take, for example, our native wood rat (*Neotoma cinara*). This bushy-tailed character goes quietly about its business, feeding on the roots, stems, leaves and seeds of plants, perhaps eating a few insects, and generally doing no more harm than pilfering the odd shiny trinket with which to adorn its nest. We call it the "trade rat" or "pack rat." *Walker's Mammals of the World* says pack rats are, "neat and solitary animals and make pleasing pets if their extreme timidity can be overcome."

Pack rats have nothing to do with rat packs or with the bad rap rats have gotten. The rats which swarm in packs through our cities and which have been called "humanity's most formidable foe," are Old World rats, imports of the genus *Rattus*. Originating in Eastern Asia, they've become commensal species—living in or near our buildings and feeding off our food, following us wherever we've settled. Commonly they're called the house rat, earth rat or alley rat, the barn or dump rat, the water, sewer, river or wharf rat. There are just two main problem species: one is the black rat (*R. rattus*) which is also called the roof or house rat. The second, and by far the worse, is the brown rat (*R. norvegicus*) known as the Norway rat.

The roof rat is a large, slim specimen with a long, naked and scaly tail, large ears and a pointed snout. A skilled climber—it can run along a strand of thin-diameter wire—it reached coastal North America in sailing ships and settled in port cities, living in the upper stories of buildings. On the west coast it has established wild populations near the forest fringe and on small islands, where it has eliminated smaller native rodents. Less aggressive than the Norway rat, the roof rat is also far less of a pest and has been largely displaced by its burly brown cousin.

The Norway rat is bigger, fiercer and more adaptable than the roof rat. A ground dweller, it's thought to have originated along stream banks in southeast Asia and gradually become commensal, spreading with the development of canals and rice paddies. An aggressive colonizer, it's now found throughout the world, living in underground tunnels or in cellars, basements or the lower floors of buildings, in garbage dumps and sewers. Blunt-nosed and thick-bodied, it can nevertheless wriggle through holes less than five centimetres wide. True to its streamside heritage, it dives and swims well.

In 1987, "huge and vicious" Norwegian sewer rats were reported swimming up drain pipes and into toilets of homes in Honolulu.

I know the feeling of dread that invading rats can inspire. When first married, my new bride, Sandy, and I rented a cute little tumble-down cottage in Richmond, just south of Vancouver. Lacking storm sewers, the area was crisscrossed with many kilometres of deep ditches running along the roadsides, and we'd sometimes spot rats swimming in the ditches. We slept on a mattress on the floor in those days, and one night I was jolted awake by Sandy. "There was a rat in your hair!" she gasped. Nonsense, I replied, a nightmare, go back to sleep. Then we both saw it: on the other side of the little bedroom, a large rat slunk out of the closet and disappeared into the kitchen. Disgust and fright shook me.

But do wild rats really merit such alarm and revulsion, or have we pictured them as vicious, dirty and despicable for reasons that have more to do with ourselves than with them? As carriers of the plague, rats became synonymous with death and disease, but as discussed in an earlier chapter, the real villain of that piece was the flea, which victimized rats the same way it did humans or any other host. Chipmunks, ground squirrels, marmots and voles can all be infected, yet none of these suffer our revulsion.

There's no question that rats spread disease. Food contaminated with their droppings can cause salmonella or food poisoning as well as leptospiral jaundice. Murine typhus fever is spread by fleas from infected rats. A bite from an infected animal can cause "rat bite fever." One research project that involved trapping over a thousand rats in Baltimore found that two-thirds of them were infected with the virus that causes haemorrhagic fever. A serious and sometimes fatal disease in Asia, the virus can cause chronic kidney disease and even kidney failure in humans.

Besides contaminating food with their urine, faeces and hair, rats steal millions of tonnes of food every year. The World Health Organization has estimated that they consume 30 million tonnes of food annually, much of that in the world's starvation areas. In other words, rats make off with enough food to feed 150 million starving people every year. In the United States alone, annual losses caused directly by rats are now estimated to approximate 1 billion dollars.

Commensal rats are omnivorous, eating a range of plant and animal matter. They'll eat everything humans do and more—including things like beeswax, paper, soap and leather. One 1975 study found the Norway rat prefers animal matter, including bird eggs, fish, mice, poultry, and young lambs and pigs. Under stress or hunger, they'll eat just about anything, and a pack will attack larger animals, even humans. *Walker's Mammals of the World* claims that "rats make direct attacks on about fourteen thousand persons annually in the United States, and occasionally inflict mortal wounds."

Like mice, rats prefer to live close to their food supply. For such world travellers, they're real stay-at-homes, often occupying a range no more than forty-five metres in diameter. Favourite haunts for urban brown rats are the alleys behind restaurants and supermarkets, where food is scattered around. In residential areas, they'll excavate burrows near homes where garbage, compost heaps, pet dishes and bird feeders provide regular and easy pickings. Hidden under shrubbery or beneath piles of bricks or boards, the burrows have numerous escape tunnels. In mild climates rats may remain outdoors all year, but in colder zones they'll move into nearby buildings for the winter, occupying barns, silos, warehouses, homes and other structures. Studies in Maryland found that an average city block is home to between 25 and 150 Norway rats, while in the countryside a farm might have between 75 and 300 of them. An old rule of thumb says that you find about the same number of rats as humans in a city. Not that you'll see them; highly secretive, rats are seldom seen in their true numbers. Another old rule says that for every rat you do spot, there will be nineteen more that you don't.

* * * * *

Life within the rat pack has proven an endless source of fascination for biologists and behaviourial psychologists. A rat colony is a strictly organized social hierarchy in which dominance and submission determine who lives where, who eats what and who gets to reproduce. Researchers have reported several different forms of social structure for the Norway rat. One U.S. Public Health Service study conducted by J.B. Calhoun demonstrated how two parallel social systems functioned in an artificial colony. In one part of the experimental enclosure, dominant males established individual territories around burrows containing a number of families. These high-ranking males excluded other males and they alone mated with females in their territory. When resident females were reproductively active, they excluded all other rats, including juveniles, from the burrow. Reproduction was regular and successful. Females collectively raised and nursed the young. In short, the territorial burrows were the essence of efficiency, discipline and harmony.

By contrast, Calhoun reported, within the same enclosure there existed a disorganized underclass in which no territories were established and whose members were condemned to a permanent submissive status. Subordinate rats, primarily males, raised in the territorial burrows were eventually forced into the disorganized area and never allowed to return to the privileged enclave. Large packs of rats formed in this disorganized area, where females in heat would be followed by dozens of males and mounted hundreds of times in one night. The stress of such mating behaviour, combined with poorly organized burrows and badly maintained nests, resulted in reproductive failure within the rat "ghetto," thus regulating the overall population in the enclosure.

Under favourable conditions, rats are prolific breeders. The breeding season lasts all year, with spring and autumn peaks. Gestation takes three weeks, and a dominant female averages five litters a year, but may raise as many as twelve. Litters average about nine naked and blind babies, but may have up to twenty-two. The young mature in about fifty-two days. Somebody has calculated that if all the offspring of a single pair of brown rats survived and bred, they could total 359 million in a year!

But the figure is meaningless, since rats are far more efficient than humans in controlling their own populations. Like other rodents, they have a number of birth-control strategies. One is "olfactory coercion," in which rodents adjust their reproductive behaviour according to scent clues in urine. In *A Natural History of Sex*, Adrian Forsyth quotes a study of Norway rats which showed female urine contains two different airborne cues—one which shortens the breeding cycle, another which suppresses estrus and lengthens the cycle. It's possible, Forsyth speculates, "that dominant and subdominant females differ in the proportion of the two chemical cues they produce."

Other strategies include spontaneous abortion. Forsyth writes that "in mammals it is a common reproductive strategy to reabsorb embryos if the mother's health or feeding success declines." Rats and other rodents "possess the ability physiologically to terminate already fertilized embryos and absorb them back into the body. A rat may get pregnant while she is still nursing offspring, but if her milking demand suddenly rises, she will reabsorb the already-implanted embryos." Females may also kill one or more members of their litters, and when overcrowding occurs within a colony, weak members may be killed, denied food or forced out of the colony. Rat fights and rat cannibalism contribute to an average life expectancy of only three months, with less than 10 per cent of rats surviving more than a year.

* * * * *

Much has been made of the rat's capacity to kill other rats, and back-porch philosophers can be counted upon to mention that of all the mammals only humans and rats kill members of their own species. "In their behaviour towards members of their own community," wrote Konrad Lorenz in *On Aggression*, rats are "the model of social virtue; but they change into horrible brutes as soon as they encounter members of any other society of their own species." As evidence, he described experiments in which researchers placed a mix of wild rats in an enclosure. "Bloody tragedies" ensued, until a dominant pair had managed to kill off all potential competitors.

Equally bloody results came from inserting a strange rat into an established colony. "What rats do when a member of a strange rat clan enters their territory or is put there by a human experimenter is one of the most horrible and repulsive things which can be observed in animals." As soon as a resident rat scents the stranger, "the information is transmitted like an electric shock through the resident rat, and at once the whole colony is alarmed by a process of mood transmission which is communicated in the brown rat by expression movements but in the house rat by a sharp, shrill, satanic cry which is taken up by all members of the tribe within earshot." Then, "with their eyes bulging from their sockets, their hair standing on end, the rats set out on the rat hunt." The strange rat is slowly torn to pieces by the residents. "Only rarely does one see an animal in such desperation and panic, so conscious of the inevitability of a terrible death, as a rat which is about to be slain by rats. It ceases to defend itself." This, observed Lorenz, is totally at odds with how a cornered rat will fight against any other foe, where even in the face of certain death it will attack, springing at the enemy, "with the shrill war cry of its species."

Lorenz went on to describe how rats engage in sustained collective aggression of one community against another. He cited the work of Steininger who, on one North Sea island, "found that the ground was divided between a small number of rat clans separated by a strip of about forty-five metres of no rat's land where fights were constantly taking place." Lorenz concluded: "It can readily be seen that the constant warfare between large neighbouring families of rats must exert a huge selection pressure in the direction of an ever-increasing ability to fight, and that a rat clan which cannot keep up in this respect must soon fall victim to extermination."

All this was welcome news during the depths of the Cold War, when the cold warriors needed a "natural" explanation for their own activities. But now it seems as though subsequent investigations have undercut the once rock-solid convictions of the "aggression instinct" proponents. For example, research has shown that a strange rat placed in an established colony will indeed be attacked. If, however, it's removed and later placed a second time into the colony, it instantly emits an ultrasonic signal of submission and the resident rats do not attack it. Similarly, urban rats have been shown to fight more than rural ones. The easy generalities about rat wars and the parallels with human behaviour are perhaps not quite so easy after all.

* * * * *

Not long ago the *New York Times* reported that, "scientists studying the rat express nothing but admiration for the rat's instincts, ingenuity and tenacity." The article quoted Dr. James Childs of Johns Hopkins Medical Centre: "They are genuine survivors; they can live under a remarkably wide range of conditions." We get an inkling of just how wide that range can be from the 1980 discovery of a rat colony thriving on the island of Runit in the South Pacific atoll of Enewetok. During the forties and fifties, a total of forty-three atomic bomb tests had totally contaminated the island with radioactivity, rendering it unsafe for human habitation for twenty-five thousand years. Yet there were the rats.

Their capacity for survival is the stuff of legend. Rats are said to leave a house which is about to fall, and sailors have an age-old superstition that rats will desert a ship before she sets sail on a voyage that will end in her loss. In *The Tempest* Shakespeare describes a ship as

> ... a rotten carcass of a butt, not rigg'd
> Nor tackle, sail, nor mast; the very rats
> Instinctively have quit it.

One of the keys to the rat's survival, writes Lorenz, is "a conservation and traditional passing-on of acquired experience." So, if a rat finds some food with which the pack is unfamiliar, it will describe to the others whether or not it should be eaten. If the food is shunned by a couple of members, no other rat will touch it. Sometimes they'll even sprinkle a poisoned bait with urine or feces. Says Lorenz, "the most astonishing fact is that knowledge of the danger of a certain bait is transmitted from generation to generation and the knowledge long outlives those individuals which first made the experience." The difficulty of successfully combatting the brown rat, he concludes, "lies chiefly in the fact that the rat operates basically with the same methods as those of man, by traditional transmission of experience and its dissemination within the close community."

Ratacake, Krumkill and Rat-Nip are three of the many rodent poisons on the market. Of the roughly two dozen registered rodenticides, one of the most popular is Warfarin, an anti-coagulant that works by reducing the animal's blood-clotting ability, resulting in death from haemorrhaging. It's touted as being less hazardous to other organisms than are the fast-acting single-dose rodenticides, which can also poison children, pets and birds. A rat has to eat Warfarin for a week or more for it to take effect. Two drawbacks: rats can learn not to accept the bait, and resistant strains of rats have evolved— the so-called "Super Rats," which are immune to the poison. In grain-handling and storage facilities, powerful fumigants are used to control rats. Traps are recommended for control of small populations, though here again a pack can learn to avoid a baited trap after it has caught one or two members.

Good sanitation is the real key to effective rat control, which means denying the rats nesting sites and food supplies. All food should be in sealed containers, garbage cans tightly lidded and compost heaps made rodent-proof. Piles of lumber or other old building supplies may cover rat burrows and should be cleared away. Beatrice Trum Hunter notes a slightly more esoteric approach: "In Vermont, a man troubled with rats succeeded in capturing the king rat and placed it in an empty milk can. The squeals scared away the remainder of the pack. The man made a recording of the distress call and played it to keep the rats away."

Sir J.G. Frazer recorded that in the Ardennes, rats could be exorcised by intoning a couple of Latin exhortations along with the words: "Male and female rats, I conjure you, by the great God, to go out of my house, out of all my habitations, and to betake yourselves to such and such a place, there to end your days." The same words were then written on pieces of paper which were folded up, one being placed under the door by which the rats were to leave, and the other on the road the rats were expected to take. This exorcism was best performed at sunrise. Similarly, it was a common belief in Ireland that rats in a field could be destroyed by rhymed incantations. Thus Ben Jonson wrote: "Rhyme them unto death, as they do Irish rats." I could certainly recommend one or two poets whose rhymes might be up to the task.

In other parts of the world, people learned to exist with, and perhaps even benefit from, their neighbourhood rats. Frazer reported that on the South Pacific island of Rarotonga, when a child's tooth was extracted, the following verse was recited:

> Big rat! Little rat!
> Here is my old tooth.
> Pray give me a new one.

The old tooth was then thrown up onto the thatched roof of the house, where rats would make their nests, in the hope that the child's new tooth would be as firm and excellent as a rat's tooth.

Let's face it: rats aren't all bad. The common laboratory rat, which is a white mutant form of the Norway rat, has contributed tremendously to medical research. Konrad Lorenz, having catalogued the savageries of rat warfare, writes: "One would hardly expect to see the development of the society which is soon built by the victorious murderers. The tolerance, the tenderness which characterizes the relation of mammal mothers to their children, extends in the case of the rats not only to the fathers but to all grandparents, uncles, aunts, cousins, and so on." Serious conflict simply doesn't occur, even when the colony numbers many dozens of animals. Food is readily shared, with smaller rats being able to push forward and take food away from larger ones with impunity. "The ceremony of friendly contact is the so-called 'creeping under,' which is performed particularly by young animals while larger animals show their sympathy for smaller ones by creeping over them."

Most of the world's rat species, *Walker's Mammals of the World* points out, occupy natural habitats and have nothing to do with humans, neither entering buildings nor pillaging agricultural areas. "Many of these species have relatively restricted ranges and habitat tolerances and some may be threatened with extinction." For example, one native species has disappeared completely from parts of Australia, apparently the victim of excessive livestock grazing. *Walker's* maintains that "the activities of a few commensal and pest species of *Rattus* have adversely affected the reputation of the entire genus and, indeed, of all mammals that contain the term rat in their vernacular names." Perhaps it could be said that, rather than humans participating in "the rat race," rats that have learned to live with us have become frantic co-participants in "the human race."

Chapter Eighteen

TOADS AND
FROGS

The Free Tenors

*B*y the middle of February, those of us who have overwintered in the forests of the Pacific Northwest are often a sodden and downcast lot. Four months of more or less relentless rain have begun to take a psychic toll. All colour seems leeched from the landscape, and sunshine lingers only as a dimming memory. One's ground of being decomposes into a soggy primal ooze. A grey melancholy clings like mist to the spirit. Uncharacteristic morbidity seeps into the collective unconscious.

Then, just as everything seems hopeless, the tree frogs begin to sing. For a night or two there may be only a few brave individuals whose preliminary croakings go almost unnoticed. But soon their numbers swell and their singing builds to a mighty crescendo, breaking over us like a tidal wave, a throaty tsunami of sound. This is not some mere haphazard croaking, some church-basement "chorus of frogs." This is a brilliant, night-shattering mad opera of the universe. Massed together in their thousands, these lusty tenors, Pavarottis of the ponds, sing every night for love! From February through May they croak and make love, and their wild rioting lifts us from our torpor and hurls us too into the luscious lasciviousness of spring!

Even as musty a place as the Canadian Club in Ottawa has heard their message. In a speech to the club in 1920, one B.K. Sandwell told the assembled swells: "Nobody but a Canadian can really appreciate frogs. As a musical instrument they do not exist in other countries, and it is as a musical instrument that they are of value in the expression of beauty in Canada."

As if to prove B.K.'s thesis that the people of other lands fail to appreciate these nightly singers' songs of love, there appeared a shocking Reuters news clip in the spring of 1991 describing how residents of the southern French town of Beddarides were complaining to town officials about singing frogs. Massed along the banks of the Vaucluse River, endangered Batrachian frogs were singing with such lusty bravado that the townspeople were becoming churlish. "There is a deafening concert from nine o'clock at night until six in the morning," complained one Thierry Sicard. "It is no longer possible to hear the television."

So much for *l'amour*. For my part, I'd rather hear the frogs. In our part of the country, we hear the songs of several frogs, a toad, and a tree frog. The frogs are the least impressive of all. Spotted, leopard and red-legged frogs, none of them has the resonance of the bullfrog, a species introduced to the coast, mercifully not yet working its mischief in our part of the woods. A true forest native, the reddish brown red-legged frog (*Rana aurora*) lives in small forest ponds and swamps. Something of a minimalist, the red-legged male's mating call is described as "a feeble, short stutter, lasting little more than a second." A tad more impressive, singing with what's described as "a high-pitched tremulous note," is the northwestern toad, also called the western or boreal toad (*Bufo boreas*). It's a short, squat creature, up to twelve centimetres long, usually coloured dark brown or reddish brown, with a conspicuous green stripe on its warty skin. Common along the coast from Alaska to California, it ranges as far east as Utah and Colorado. Almost deafening in comparison is the song of the Pacific tree frog, which ranges from Vancouver Island to Baja. These little chaps, less than five centimetres long and delicate in appearance, can make an astounding amount of noise for creatures their size. At the height of their mating frenzy in March, you can hear the big congregations from one or two kilometres away.

Pacific tree frogs (*Hyla regilla*) are coloured everything from bronze to brilliant green, but are easily distinguished by tiny discs on each of their toes and by a dark band which stretches like a mischievous grin from nostril to shoulder. The handbooks used to call them tree toads, but in recent years they've been reclassified as tree frogs. Is there a difference? Not really, say herpetologists. Originally the two terms applied only to European species: frogs all belonged to the genus *Rana* and toads to the genus *Bufo*. Generally toads were thought of as being warty-skinned, squat, short-legged, more at home on land than in water and possessing poisonous glands; frogs were characterized as smooth-skinned, long-legged, more aquatic than terrestrial and lacking any toxins. However, only a minority of the twenty-eight hundred species of toads, frogs and tree frogs found worldwide belong to these two genera, and the reasons why they're called one name or the other are often spurious. In their handbook on B.C. amphibians, David Green and Wayne Campbell from the Royal B.C. Museum tells us that "there are Spadefoot Toads, Chorus Frogs, Narrow-mouthed Toads, Clawed Frogs, Fire-bellied Toads and Poison-dart Frogs, none of which is particularly closely related to ranid frogs, bufonid toads or to each other. No wonder scientists just use their Latin names."

* * * * *

The principal difference, it would seem, lies in the prejudiced eye of the beholder. In European tradition, toads have all sorts of horrid connotations that we don't attach to frogs. Theologically speaking, for instance, toads have had a very poor reputation over the years, while frogs have been accused of nothing worse than the biblical plague, when "the frogs came up, and covered the land of Egypt." Toads are another matter: in the sacrilegious Black Mass of the Middle Ages, satanists reportedly used toads in profane mockery of the Christian Eucharist. In mediaeval times, toads—along with spiders, snakes, bats and other creatures—were believed to be the demonic servants of witches. Thus the soothsaying witches in *Macbeth* chant of the "swetter'd venom" of the toad. And when there's a witch to be seen, can the indefatigable Saint Patrick be far behind? The absence of toads in Ireland is also credited to the good saint who "cleansed the island of all vermin by his malediction."

Centuries later another priest, Louis Hennepin, exploring the wilds of the New World with LaSalle, reported that many Indian tribes "have infallible remedies against the Poison of Toads, Rattlesnakes and other dangerous Animals." Ironically, the most poisonous amphibian is a Latin American frog, not a toad, and yet toads retain an unshakable reputation for being ugly and venomous. A toadstool is a poisonous and unwholesome sort of mushroom; touching a toad will give you warts. Ancient toad-stones were believed to have come from toads and were used in amulets as protection against poisons. In 1597, Thomas Lupton wrote that "a toadstone touching any part envenomed, hurt, or stung with rat, spider, wasp, or any other venomous beast, ceaseth the pain or swelling thereof."

We use the word toady to describe a fawning flatterer, parasite or sycophant, and the old epithet toad-eater had the same vile connotation. The *Oxford* dictionary says that toad-eater originally derives from "the attendant of a charlatan, employed to eat toads (held to be poisonous) to enable his master to exhibit his skill in expelling poison." Thus Tom Brown's satirical couplet,

> Be the most scorn'd Jack-Pudding of the pack,
> And turn toad-eater to some foreign quack.

Saddled, unfairly, with this age-old and unfortunate heritage, toads were described in the first formal zoological treatise in English, by Thomas Penant, in 1776, as "the most deformed and hideous of all animals—objects of detestation." Even today *Oxford* gives as a second definition of toad, "a type of anything hateful or loathsome."

* * * * *

Hanging on my study wall beside my desk is a wonderful *New Yorker* cartoon by "Anthony" which depicts, in six panels, a small creature emerging timidly from a vast ocean onto a beach. The animal's little bubble-thoughts form an image of the apex of evolution: an executive male in suit and tie, trudging to the office, briefcase in hand. Horrified at what it might be beginning, the little creature turns about and sadly slinks back into the ocean.

Perhaps if they'd imagined what lay ahead, the earliest amphibians would have retreated too, but they didn't. Instead they emerged, 350 million years or more ago, as the primal ancestors, not just of toads and frogs, but of ourselves as well. It was they who made the gigantic evolutionary leap from living in water with gills to living on land and breathing through lungs. For over 100 million years, writes British amphibiaphile Trevor Beebee in *Toads and Frogs*, these ancestral amphibians, looking like primitive salamanders, were a dominant form of terrestrial life. Their first real predators appeared with the evolution of reptiles some 200 million years ago, which was also about the time that identifiable frogs began appearing.

Occupying a unique evolutionary niche between fish and reptiles, amphibians are of tremendous interest to biologists. Frogs and toads particularly have developed what Green and Campbell call "astounding variation"— from massive to miniature, timid to aggressive, cunningly camouflaged to dazzlingly coloured. The Goliath frog of the Cameroons, for example, can grow to almost thirty centimetres long, while the infamous marine toad of tropical South America approaches twenty-five centimetres in length and 1.5 kilograms in weight. Other species are so tiny as to be barely noticeable. Common toads have been found at an elevation of eight thousand metres in the Himalayas as well as over three hundred metres beneath the earth's surface in coal mines. What is most remarkable and wonderful about them all is that every spring, in their procreation, they re-enact that ancient primal emergence out of the waters and onto land.

Generally speaking, frogs tend to live close by their home ponds, while toads and tree frogs are more prone to wandering off elsewhere. But as the mating season approaches, toads and tree frogs begin the great migration, usually to the ponds of their birth. How they find a particular pond remains a mystery. Some researchers believe that the animals can detect a distinctive scent produced by the plants of the pond, though there are many stories of toads continuing to migrate to a pond even after it has been filled in for several years. In his book *Animal Reproduction*, Philip Street writes that "homing is beyond dispute, and toads are generally more successful than frogs at it." Frogs, for example, will quickly colonize a new garden pond, while toads are far more particular, passing by any number of apparently appropriate ponds in search of the right one.

Writing about common English toads, Trevor Beebee says they have "three ways of getting about—swimming, a leisurely crawl or a pathetic hop." Their springtime migrations become an extended crawl, sometimes of 1.6 kilometres or more. They can cover up to 150 metres an hour, but need to pause for a rest every few steps, which can be a fatal mistake on roadways. Along our island roads we see the little tree frogs in migration, though they hop rather than crawl. Travelling by night in this slow but steady pilgrimage, toads overcome tremendous obstacles to get where they're going, including climbing over walls and fences. Sometimes the males make this trek ahead of the females, but if by chance a male and female meet en route, the male will immediately hop onto her back, grip her in the mating embrace known as amplexus, where he locks his arms under her armpits and across her chest, and refuse to be dislodged. Some males develop special swellings on their forefingers, called nuptial pods, which help them clasp the female unshakably. Thus seized upon, the female is then forced to carry him the entire distance to the pond, even over the worst obstacles.

While frogs often straggle into their mating ritual in dribs and drabs, toads may arrive en masse in a grand "splashdown" involving the entire breeding colony. And then the music begins! The songs are produced, primarily in males, by passing air over vocal chords in the larynx, and in many species are amplified through vocal sacs which act as resonators. With mouth and nostrils shut, the singer passes air back and forth between his lungs and vocal sac, turning his whole body into a little "boom box." In some species the vocal sac can swell to three times the size of the singer's head. Others, such as the red-legged frog, sing their songs underwater.

A pond at mating time may sound like a garage band on acid, but is, in fact, a carefully orchestrated performance. Each species has its own distinctive song with mating behaviour to match. Many true frogs, write Campbell and Green, "set up territories in breeding ponds and actively defend them with calls, threatening displays or wrestling with intruders." Pacific tree frogs also tend to sit in one place, carefully spaced between neighbours. Their songs are both advertisement and assembly calls, designed to attract females, and territorial calls, warning other males to keep their distance. With some of the toad species, things are a lot more chaotic. "When male toads are in breeding spirit," Campbell and Green continue, "they may be aggressive and attempt to grab just about anything that looks like a conspecific female, including other males." Even fish can find themselves embraced! In this wild press of flesh, calls can serve to advertise one's gender. Studies of the common toad in Europe have shown that the pitch of the male's croak is related to his body size and that the larger he is, the deeper his croak. When a male has grasped a female in amplexus, often other males will attempt to dislodge him. He responds by croaking, and if his croak is sufficiently deep, the attackers realize that he's too big to be dislodged.

Sometimes the mating songs are only one pure note, given over and over. Our northwestern toad is one of these monotones, repeating its single note in rapid succession. Other species tend to pause between each of their notes. The Pacific tree frog's call consists of two notes, like a bullfrog's only not so basso profundo, and written as "wreck it!" This croak is repeated every few seconds, but not at random. Rather, the singers take turns and space their calls between those of their neighbours, often forming duets and other antiphonal effects. The caller switches to a single-note song of warning if another singer draws too close, but often the female he has worked so hard to entice will actually have to touch him before he realizes she's there for him. In some species of tree frog, certain males don't sing at all, but sit quietly near a good singer and attempt to catch approaching females. Some frogs simply pump up the volume of their territorial calls if a rival begins calling too close at hand. Thus, as Trevor Beebee writes, the mating pond "isn't the random cacophony of sound it at first seems, but a den of intrigue and competition for the only success that really matters—fathering the next generation."

* * * * *

After all this loud licentiousness comes the familiar laying and fertilizing of spawn, hatching of eggs, development of tadpoles and the wonderful transformations of metamorphosis. But horrendous mortality stalks each stage. Freezing or fungal infections may kill the spawn. Small ponds may dry up, leaving the eggs desiccated. Tadpoles are attacked and eaten by fierce water beetles, dragonfly nymphs, fish, newts, shrews and birds. In the twelve weeks that it takes tadpoles to achieve metamorphosis, writes Beebee, less than 5 per cent of their number will survive. As the little toadlets and froglets emerge in midsummer, birds of prey attack them at pond's edge. Those that escape these hazards spend a life on the run from fish and snakes, from gulls, herons, owls and crows, from otters, rats, foxes and cats.

Perhaps the most gruesome death of all comes from the repulsive toad fly, a greenbottle which parasitizes common toads in Europe. The fly lays her eggs on the toad's back, and after hatching the maggots crawl into their host's nostrils, mouth and eyes. The maggots eventually eat their way through the toad's innards, leaving literally only skin and bones, before pupating.

Stalked by killers at every turn, writes Beebee, toadlets will suffer a further 95 per cent mortality before reaching sexual maturity in two or three years: of two thousand eggs that hatch, perhaps five adults will survive long enough to mate.

A major problem for toads and frogs is that they may simply fail to see approaching predators. As every child who has hunted frogs in a ditch knows, you can easily catch them by hand merely by approaching slowly and stealthily. Though they have large and brilliantly coloured eyes, capable of seeing in almost 360 degrees, toads and frogs have a highly specialized kind of sight designed to detect the movement of small insect prey against a background. It's incredibly good for this purpose, and the eye-to-tongue coordination is astounding; one expert called the flick of the tongue "one of the most spectacular muscle actions in nature." But, strangely, the eyes see only what moves or changes, and so remain blind to the stealthy approach of a snake or the painstakingly slow advance of a predatory heron.

Naked to the world, without any protective scales, fur or feathers, frogs and toads have developed really versatile skins. Camouflage is their first line of defence, and a good one. Skin pigments, warts and tubercles help the creature blend into its background. Some species, like the Pacific tree frog, are able to change colour by expanding or contracting the pigment cells of their skins. However, the colour changes of the Pacific tree frog don't relate to the colour of the background, but rather to changes in temperature and humidity. On cold, wet days the little frog's colour is much darker than the almost-fluorescent greens and bronzes of bright days. These weather changes also stimulate the tree frog's "rain song"—a wooden, rasping and dispirited sound that we hear throughout the summer months. Not quite chameleon-like, still they're hard to spot as they blend into the colours of leaves and bark. With a tiny suction cup at the tip of each toe and finger, these little fellows are beautifully adapted for life in trees and shrubs. The suction pads adhere to surfaces through what researchers call "capillary adhesion": a fluid on the frog's skin creates a surface tension that permits it to climb up and even fall asleep on a vertical sheet of glass without falling off!

Besides camouflage, the skin plays other important roles—it's part of the animal's breathing apparatus and allows frogs to spend the winter months buried in mud at the bottom of frozen ponds. In summertime, special glands help keep the skin damp and supple, and in place of drinking, frogs and toads simply absorb moisture through their skins by sitting in damp places. And as a last line of defence, toads especially can secrete toxins which make them instantly distasteful to many predators. There are two types of repellents produced by warts or glands, which are most numerous behind the toad's eyes. Toad poisons include four active substances—amines, alkaloids, steroids and steroid esters—which can constrict blood vessels, raise blood pressure, and excite muscle spasms. Some bufotoxins, writes Beebee, are as poisonous as strychnine, while at least one toad alkaloid is a potent hallucinogen.

It's this latter quality that has generated the recent fad of getting high by "toad-licking." As early as 1970, the U.S. federal Drug Enforcement Agency, ever vigilant against mind-altering substances, banned bufotinine as a contraband drug. Bufotinine was first found in the Sonoran desert toad of the southwest, and, said one government official, "If you tried to lick this toad it would be a felonious act." Predictably, interest in toad-licking zoomed, and has generated a sizeable cult following.

* * * * *

These days, the dangers to stoned toad-lickers are far less than to toads and frogs themselves. Within the last two years, scientists have been sounding a frightening alarm: toad and frog populations worldwide are crashing at an astounding rate. Fluctuations in amphibian cycles are quite normal, but what we're seeing now, say the experts, is both extraordinary and appalling. Whole populations are simply disappearing. Typical is the case of the yellow-legged frog in California's High Sierra. A recent *New York Times* article describes how these frogs, which were plentiful in the seventies, have disappeared completely from dozens of ponds.

"Here is a whole class of animals that seems to be in decline from North America to South America, Africa to Asia, and we don't know why," Oregon State University's Andrew Blaustein recently told the press. Perhaps pesticides drifting up from the over-sprayed fields of California's Central Valley are to blame for the disappearance of the yellowlegs in the Sierras. In other areas, agricultural pesticides and chemical fertilizers are blamed for poisoning frogs living in drainage ditches and collection ponds. Similarly, toxic rain and snow, particularly acid rain, are suspected to be at least part of the problem. Dependent upon both land and water, amphibians are exposed to the contaminants of both. Some experts speculate that ozone depletion may be increasing the amount of ultraviolet light and adversely affecting eggs exposed to sunlight in shallow ponds.

Human consumption of frog legs causes problems in some areas. About 200 million frogs are killed each year to provide the delicacy, and three-quarters of that total come from the Far East. Recognizing the critical role of frogs in insect control, India banned the export of frog legs in 1987, leaving Bangladesh and Indonesia as the major exporters, though rumours continue to circulate about illegal export of frogs from India. In other places, the misguided introduction of alien species has had devastating consequences for native toads and frogs. The classic case is of the marine toad (*Bufo marinus*), which is now widespread throughout Africa, Asia and Australia where it has reached plague proportions. Introduced as an insect predator, these lugubrious giants have proven very efficient at decimating native amphibian stocks as well. Less dramatically, ill-advised attempts to establish "frog farms" have led to the release of American bullfrogs (*Rana catesbeiana*) in the Pacific Northwest and elsewhere. Large and aggressive, they're wreaking havoc on native species too.

Perhaps most important of all on a worldwide basis is the destruction of habitat, none more disastrous than the burning of tropical rainforests which are home to a fabulous diversity of amphibians. In our country, clearcut logging has done untold damage, eliminating old-growth and mixed species forests, choking creeks and ponds with silt and debris, and destroying wetlands by running skidders and other heavy equipment through them. In B.C., "something on a very large scale is certainly happening to the leopard frog," a spokesperson of the Royal B.C. Museum told the local press. "In Washington and Oregon, spotted frogs have been disappearing." Elsewhere, agribusiness practices have drained swamps, ponds and wetlands, and urban sprawl has done the same. Roadways cut across migration routes and untold thousands of toads and tree frogs are squashed beneath the tires of hurtling automobiles.

But most confounding of all to the experts is the dramatic decline now documented in populations living in virtually untouched wilderness. David Wake, a biologist at the University of California at Berkeley, recently told the *New York Times* that frogs and toads "are disappearing from nature preserves, in the most pristine sites of Costa Rica, Brazil, Yosemite, Sequoia and Isle Royale National parks. Meadows where frogs were as thick as flies are now silent." Wake and other experts think the mass die-offs should be giving us an unmistakable message—that ecosystems may be starting to collapse, and amphibians are acting as "indicator species," sounding an alarm that we ignore at our peril.

Where I live that wholesale population crash has not yet occurred, at least not among the tree frogs whose songs are, if anything, more boisterous in recent years. The natives of this coast held these mighty singers in great regard, and the Frog Crest is an ancient and venerable one. In foggy weather, the singing of the frogs was said to guide mariners through the treacherous waters of the coast, and totemic frogs gave protection to the poles upon which they were prominently displayed. It was said that Haida shamans might acquire their healing powers through the frog, which symbolized wisdom, and that two large frogs stood as guards to the Thunderbird kingdom, croaking a warning at the approach of strangers.

Today they croak another warning, one that signals the approach of a mortal danger to us all. I hate to even think of a spring silent of these singers, for it would be no spring at all. I hate to think of the consequences of not heeding their song.

Chapter Nineteen

WASPS
The Social Terrorists

I have every reason to fear and loathe yellowjackets, for they've broken my heart, cost me the affection of my first true love and taught me the painful and perverse lesson that "big boys don't cry." My brothers and I and some other neighbourhood kids were having a rotten apple fight one afternoon in the backyard of Diana Maclean's house. Eleven years old, I was madly in love with Diana, but locked in rivalry for her affection with my older brother Gerry. A huge old apple tree grew in Diana's backyard, and on that fateful day we stood on opposite sides of it, pelting one another with rotten windfalls. My brother hurled an apple high and wide, splattering into the tree. In an instant I was aswarm with angry wasps. Attacking viciously, they stung me on my neck and arms and face. I screamed and flailed at them and ran for my life! And then I did what's unforgivable in any knight errant—I cried. I sobbed from the pain and terror of the attack, but my brothers mocked me as a crybaby and, far worse, Diana scorned me, rejected me forever as unworthy of her favours!

Totally humiliated at the time, in hindsight I find my response entirely appropriate, because a mass attack of wasps is a terrifying experience. They buzz violently about your face, and strike with numbing pain in rapid-fire staccato flashes. The stabs of pain and whirling terror disorient you in seconds. You're outnumbered, hopelessly overmatched, you want to cry "uncle!" but the swarm continues its vicious assault, gets tangled in your hair, trapped under your clothing, buzzing fiercely and stinging you into submission.

"An ingeniously designed biochemical weapon of terror," is how Adrian Forsyth described bee and wasp stings in a *Harrowsmith* magazine article. The stinger itself is a sheathed dagger, a sophisticated piece of equipment hidden within the insect's body. Originally an ovipositor, designed for laying eggs, it has evolved into a lethal weapon because the glands attached to it hold one of the deadliest venoms known in nature. Though it happens with lightning speed, the act of stinging is an intricate one, precise as microsurgery. The attacking insect first takes a firm grip with its legs, then lifts several plates at the rear of its abdomen, thus exposing the apparatus of attack. With a downward thrust of its abdomen, it drives the stinger into flesh. The weapon itself has a central stylet sheathed in two barbed lancets. As it's driven down, the barbed teeth grip their way deeper and deeper into the flesh. Simultaneously, a contraction of muscles around the poison sacs squeezes their venom through the conduit formed by the stylet.

The intense flash of pain you feel is not from the stinger but from the venom, which packs an instant wallop. People vary greatly in their reactions to stings, but typically there's an acute burning sensation right at the point of stinging, quickly followed by swelling and extreme itchiness, which may spread across nearby parts of the body. Forsyth says the venom is an ingenious blend of substances, including molecules which stimulate severe pain, transmit nerve impulses, regulate moods, break down body tissues and accompany allergic reactions. "Biochemists," he writes, "praise these venoms as the most elegant toxic secretions yet evolved."

Even though injected in microscopic amounts, these powerful concoctions can trigger systemic reactions ranging from quite mild to life-threatening. Milder symptoms may include fever, headache, malaise, itching, throat constriction, stomach cramps, nausea, vomiting and dizziness. Writing in the journal of the American Medical Association, C.A. Frazier adds, "Severe systemic reactions may include labored breathing; difficulty in swallowing; hoarseness or thickened speech; weakness; confusion and a feeling of impending disaster." A shock or anaphylactic reaction, he continues, "might also include lowered blood pressure, cyanosis, collapse, incontinence and unconsciousness." Untreated, the sting can be fatal, and sometimes delayed reactions, several hours after stinging, can also result in death.

We don't know precisely how many people die each year from bee and wasp stings, but it's certainly more than from black widow bites or rabid bats. Other than disease-carriers like the flea, no insect kills more people; I read one estimate that put the worldwide total at over forty thousand per year! That seems high: a 1973 report of the Insect Sting Subcommittee of the American Academy of Allergy reported more than four hundred fatalities over a ten-year period. The report said that one hundred autopsies revealed that sixty-nine sting victims died from respiratory failure, twelve from shock and another dozen from "vascular involvements," seven from neurological problems and two from bacterial infections. A U.S. Department of Agriculture handbook estimates "yellowjackets probably cause far more than the reported fifteen to twenty deaths per year in the U.S., because the symptoms of allergic reaction to stinging are extremely similar to those of heart attacks and are undoubtedly often mistaken as such."

All true wasps have a sting, but of the many thousands of species found worldwide, almost all are solitary and inoffensive creatures, doing far more good than harm. Entomologists have a mind-numbing categorization of them all, with enough genera and species, families and subfamilies to make your head spin. There are, for example, thousands of species just of parasitic wasps, many of them infinitesimal, which lay their eggs on or in the body of a host insect. Some use their sting to paralyze the host insect, others don't. Hosts include many of humankind's worst "pests"—the gypsy moth, coddling moth, various leaf rollers, caterpillars, cutworms, aphids and beetles. Many parasitic wasps are so specialized they employ only one host species.

Other species, like the solitary potter and digger wasps, lay their eggs, not in a host animal, but in individual chambers or cells constructed from mud. Here the mother wasp provides her unborn offspring with a dead or paralyzed body of an insect or other small invertebrate. She uses her stinger to kill or paralyze this prey, then seals it in the cell with her egg, so that her larva can feed upon the insect between hatching and pupation. "In some species," writes Philip Street in *Animal Reproduction*, "the female is so skilful she can penetrate the main nerve ganglion of the victim with her sting," paralyzing but not killing the insect, so that it remains alive as fresh food for her own hatching offspring. In other species the venom kills the insect, but also contains an antiseptic which prevents the corpse from decaying before the larva can hatch and eat it.

* * * * *

But in social wasps—the swarming aggressors we've come to fear—the sting has evolved into something else entirely: not a tool of procreation, but rather a weapon with which to defend the colony. Long ago certain solitary wasps began evolving into social creatures that lived in colonies whose members developed highly specialized roles, tremendously modified reproductive patterns, and a building technology capable of quickly accommodating thousands of new members. Though these elaborate nest-builders are what we have in mind when we think of wasps, they are only a tiny minority of all wasp species, the exception rather than the rule.

In North America, there are seventeen different species of social wasps; one of them is an introduced hornet on the East Coast, the other sixteen are natives belonging to two subfamilies of Vespidae wasps. Americans generally use the term "yellowjacket" to include all sixteen native species, though technically it refers to only one of the subfamilies. We'll do the same here. True to the name, most of these species are coloured yellow and black, though a few are white and black, and some are marked with red. Experts can distinguish even very similar-looking species by the distinctive colour patterns on the large rear body section, which is called the gaster.

Though some species construct their nests in cavities in the ground or in logs, while others hang their nests in the open air, all yellowjackets build nests out of paper carton. Writing in *Natural History* magazine, Glasgow University zoologist Michael Hansell says, "I think the ability of some wasps to manufacture good, strong paper contributed significantly to the evolution of social behaviour in these species." No solitary wasps possess this skill, but in the social species, paper-making and nest construction are highly developed. The worker wasps, always sterile females, rasp bits of wood fibre from nearby wood surfaces and form the cellulose into tiny pellets of pulp which they carry in their mouth parts back to the nest. There the worker chews the wood fibres with its sturdy jaws, mixing in saliva, and gradually spreads the resulting pulp into a thin sheet. Michael Hansell observes that "thinness of walls and abundance of saliva sometimes gives the paper a glistening, transparent quality." He has calculated that one yellowjacket nest in his collection must have required a minimum of 200,000 loads of pulp to build! But for the burgeoning colony, the result is worth the work—a remarkably thin and light, yet tough, protective sheath.

Capable of holding many times its own weight, wasp-paper gives the colony a wider range of options to ensure reproductive success. Hansell observes that "one principle of paper architecture is fundamental: it hangs, and anything that can be hung has an improved chance of being placed where walking predators, at least, are not likely to set foot."

Wasp colonies, unlike perennial bee hives, are founded and built, flourish and collapse within a single growing season. The colony is established in the spring when a fertile queen emerges from her winter sleep and constructs a simple nest, either hanging from a branch or other raised surface, or in some underground crevice, such as an abandoned mouse hole. She gathers and masticates a bit of wood pulp, fashioning a paper disc from which a thin stalks hangs. Off this stalk she suspends a horizontal layer of several dozen hexagonal cells in each of which she deposits a fertile egg. A paper envelope covers the cells. In due course, her first-born emerge and immediately set to work feeding their queen and enlarging the colony. The first comb is extended laterally, then a second comb is begun beneath it, hung by little tension struts. In subterranean nests, workers excavate the hole, removing soil to make room for the expanding colony. By the height of summer, a large colony might have eight or more horizontal combs, one below the other, containing up to ten thousand brood cells, with three thousand or more active workers feeding the queen and her larvae.

As colonies evolved over time from similar solitary beginnings into vast and complex communities, the purposes and architecture of nests evolved as well. In temperate zones, writes Hansell, "climate was probably more important than predation in shaping the architecture of the nest." The best way to conserve precious heat within the nest is to reduce its outside surface area, and stacked combs within a sphere accomplish this admirably. For the same reason, the outer wall, rather than being formed from a single thick layer of paper carton, is made from many thin layers of paper, like phylo pastry, with pockets of insulating air trapped between them. As nests continued to grow bigger, Hansell writes, safety from predators relied less upon concealment than upon mounting a stout defense—again, the spherical nest with limited access presents the smallest possible defendable surface.

And it's here that the worst possible interface between humans and yellowjackets happens. Stirring up a hornet's nest has become proverbial for getting into big trouble, and for good reason—when a nest is disturbed, its defenders attack the intruder with suicidal zeal. For all the nest's female warriors know, the disturbance may be from a predatory skunk, coyote, raccoon or bear, any of which may tear a nest apart and feed upon its larvae. Their attack response is instant and unrelenting, even though stinging the intruder may cost a defender its life. Most social wasps have a barbed stinger that remains lodged in the skin of a victim, rupturing the wasp's gaster. Even with her life already ebbing, she continues to harass and attack the intruder.

This past summer I was working in the woods, limbing a big fir tree I'd just felled, when by chance I saw a telltale swarm of blackjackets buzzing angrily. These are *Vespula consobrina*, a black and white woodland wasp which nests in hollow logs and is the bane of loggers in the Pacific Northwest. I was not within seven metres of the nest, and yet the vibration of the falling tree had triggered the battle instincts of the nest defenders. On another occasion, the reverse happened: I was about to buck up a tree that had fallen the previous winter. I bent down to look under it for obstructions and found myself face to face with a nest the size of a basketball! Fortunately, I hadn't touched the log, and though I was close at hand, the workers were going quietly about their business mere centimetres from my face. I tiptoed away with the exaggerated stealth of a cartoon villain. Another time, just down the road from us in a pile of logging slash, a similar huge nest hung from a lovely big fir log. All summer long, unwary firewood cutters, who hadn't heard otherwise, tried putting chain saw to that prime log, but no one got a single stick of wood from it, so relentless were the wasps in defense of their home, and several cutters got multiple stings for their trouble.

* * * * *

Throughout the summer months, the queen continues laying fertilized eggs, which produce new worker wasps for her colony. Yellowjackets do not produce or store honey the way bees do, but feed the larvae mostly meat, from the flesh of insects or carrion. I've watched a horde of them quickly dismember a whole mouse carcass and carry it off to their nest. The workers masticate these chunks of meat before feeding them to the larvae, and feed themselves on the juices produced in chewing. Scavenging in large numbers for fruit and carrion again brings yellowjackets into conflict with humans. One of the worst pests for this is the common yellowjacket (*Vespula vulgaris*), a ground-nesting species that specializes in scavenging from animal carcasses, garbage cans and picnic tables. Another ground-nesting species, the western yellowjacket (*V. pensylvanicus*) is the primary wasp pest on the West Coast, terrorizing picnickers, fruit pickers and loggers. Every four or five years this species undergoes a population explosion, and in 1973 a huge population outbreak of both common and western yellowjackets traumatized people in Washington and Oregon—campgrounds were deserted, logging camps and sawmills shut, food processing plants abandoned.

Fruit pickers often suffer the worst of both worlds, having to deal with both nests and scavenging wasps. Yellowjackets love ripe grapes, peaches, plums, apples and other fruit. They'll chew their way right inside, and an unwary picker may pluck what looks from one side to be a sound piece of fruit, only to find half a dozen angry yellowjackets feeding inside. Compounding the problem, as I unhappily discovered beneath Diana Maclean's apple tree, several aerial-nesting species often build their nests in fruit trees. One of the most common is the "bald-faced hornet" (*D. maculata*), a large black-and-white wasp which builds particularly big and durable nests. These wasps are not overly aggressive, unless their nests are directly disturbed. The aerial yellowjacket (*D. arenaria*), one of the commonest wasps across the country, also builds large hanging nests and is absolutely fierce in defense of them. These are the annoying characters that often buzz repeatedly around your head when you're walking in the woods; they're not attacking, but foraging for flies normally found near large animals like ourselves.

By late summer, when foraging wasps are at their worst, the queen changes her egg-laying habits, and the whole colony begins working towards next year's nests, which none of them will see. The queen lays some unfertilized eggs now which will hatch into males whose sole purpose is to fertilize the new queens. As Michael Hansell puts it, "In all matters other than sex, male wasps are quite useless." At the same time, workers construct some extra-large brood cells into which the queen deposits fertilized eggs, just like those she has laid to produce the sterile female workers. But the larvae of these chosen eggs are fed a special diet rich in proteins, and they emerge from pupation not as sterile workers but as the next generation of queens. They leave the nest, take wing in a nuptial flight, are fertilized by the males, and then seek out sheltered crevices in which to hibernate. The rest of the colony is doomed to death with the onset of winter. As food supplies dwindle, workers will pull larvae from their cells and feed them to other larvae or discard them. In these final weeks of life, torpid workers of some species are more likely to sting than they were earlier.

In a few atypical instances, in warmer climates and sheltered locations, whole populations do survive and become perennial colonies with multiple queens. Their nests can grow to enormous size—one found in New Zealand measured 3.7 metres long, weighed about 450 kilograms and contained several million brood cells. Another one discovered in California had twenty-two functioning queens, twenty-one comb levels and over sixty thousand workers!

* * * * *

Even when yellowjackets are out in force, you can reduce your chances of being stung. Sealing all garbage and foodstuffs is a good start. A U.S. Department of Agriculture handbook lists a number of other preventive steps: avoid using perfumes, hair sprays, tanning lotions and cosmetics, as these compounds often attract yellowjackets, which have poor eyesight but a keen sense of smell; wear light-coloured clothing such as whites and tans; avoid walking barefoot; be particularly watchful when gardening or mowing lawns; and avoid outdoor cooking during the height of yellowjacket season. Above all, if a wasp gets close, don't panic! Calmly brush a yellowjacket away, don't swat at it wildly. If a wasp gets in your car while you're driving, the handbook recommends pulling over and removing the intruder, rather than flailing wildly at it in the fast lane. The one exception to this cool and calm approach is when you've accidently hit a nest and aroused its defenders. Then, as Adrian Forsyth says, the natural response—to run for your life—is the right one!

For most people, a yellowjacket sting is a painful but short-lived experience. Dealt with quickly, the pain can be greatly reduced. Firstly, if there's a stinger still impaled in your skin, you should remove it right away.

But don't try to grab it and pull, warns Forsyth, because you may end up squeezing additional venom into the sting from venom sacs still attached to the stinger. Better, he suggests, to remove the stinger by drawing a knife blade or fingernail across your skin. Pain can be reduced by immediately applying any one of several substances. Some people maintain that common mud's the best thing to plaster on. We've had excellent results with just a few drops of ordinary bleach as well as with baking soda dampened with a little water. A wad of wet chewing tobacco is another time-tested poultice. Household ammonia is supposed to work well too, as is kitchen meat tenderizer. If all else is lacking, a piece of ice applied to the sting will prevent the toxin from spreading and dull the pain somewhat.

Things are more complicated for the minority of people who suffer allergic reactions to wasp venom. In extreme cases, people have to carry emergency kits which contain a preloaded syringe of epinephrine for immediate injection after being stung. Kits may also contain antihistamine and phenobarbital tablets or an inhaler charged with a bronchodilator. Long-term relief for extreme allergic responses entails a lengthy desensitization program involving venom injections.

* * * * *

Gardening magazines and seed catalogues offer all kinds of contraptions for making your yard and garden "wasp free," but that objective is neither as simple nor as desirable as it sounds. There are various kinds of traps that foraging yellowjacket workers can get into but not out of. These offer minor abatement, but do nothing about the breeding colony. For those determined to destroy a nest, there are poison baits on the market, generally encapsulated diazinon. Mixed with tuna-fish cat food, these have proven successful against meat scavengers like the western yellowjacket, but, again, they involve releasing toxins into the environment and extreme care must be taken not to poison birds or small mammals.

Similarly, one can buy "wasp bomb" aerosol products, which squirt an instant-knockdown insecticide three metres or more at aerial-nesting colonies. Forsyth recommends a commercial pyrethrum spray, a natural insecticide that biodegrades rapidly. For underground colonies, he says, simply wait until evening and dump a load of soil over the entrance, first making sure that you've got the right and only entrance. Perhaps the toughest nests to deal with are those inside a house wall. You can't just block the entrance hole because, warns the Department of Agriculture handbook, yellowjackets may escape by chewing holes through your interior walls. Exterminators deal with the problem by inserting an aerosol nozzle into the hole, spraying, and then sealing the entrance with pesticide-soaked steel wool. This is not something an amateur should attempt, particularly not without protective clothing. Equally risky is the old method of smoking out an aerial nest with a smouldering oily rag on a stick.

The first rule of thumb in nest-elimination is to do it, if possible, before the nest gets too large. A couple of years ago, we had a half-dozen yellowjacket nests established along our rose arbor. Every time we touched the roses—to smell them or prune them or anything else—out sprang the wasps looking for a fight. It just wouldn't do. So one evening at dusk I climbed into my battle fatigues: first I put on a cumbersome jet fuel suit, a single-piece heavy plastic coverall I got from an army-surplus store. Then a thick hood, mask and goggles, boots and thick leather gloves. I waddled up to the arbor, looking like some overweight thrift-store Darth Vader. Impervious to any sting, I simply flicked the nests into a large plastic bucket containing a few inches of water and a splash of gasoline.

But this kind of warfare should only be undertaken when absolutely necessary, for wasps play a tremendously important role in controlling insect populations and in recycling carrion. A ''wasp free'' garden is a sick garden. As Philip Street concludes, ''The wasp, then, despite its sometimes irritating presence, is certainly a beneficial insect; and killing queen wasps in the spring, and later destroying wasp nests (except when they are very near to buildings) is a very short-sighted policy.'' Our approach is, whenever possible, to live and let live—one year we had a big nest right under the house, but we never bothered it and the busy workers never bothered us.

Interestingly, wasps are yet one more of our despised species that superstition once associated with witchcraft. A mischievous witch was believed able to leave her body and assume the form of a wasp in order to go about on various evil errands. From a totally different tradition, it's said that certain native bands along the West Coast would have their warriors prepare for battle by rubbing their faces with the ashes of burned wasps so that the men might fight as fiercely as the insects. Things might have turned out differently for me if I'd done the same before going into battle for the love of Diana Maclean under her apple tree.

Chapter Twenty

WOOD DECOMPOSERS

Dealing with the Debris

*T*he sky was a clear bright blue one June morning when I glanced out my kitchen window and saw that the huge tree was dead. For years I'd watched its slow decline, an old-growth Douglas-fir, one of the forest's last remaining giants. Each year a few more of its topmost branches would have shed their needles and twigs, death advancing downwards, leaving a gnarled top of dead limbs crooked like arthritic fingers against the sky. Now, with a shock, I saw that the last thin clusters of green needles had completely browned off and death had irretrievably claimed the tree.

Twenty years ago we'd bought this little patch of Gulf Island woodland, in part for the huge ancient trees growing in rich earth along a tiny seasonal brook. The veteran Douglas-fir in the southeast corner was the biggest of them all—about two metres in diameter at the butt, a massive columnar trunk seventy metres straight up and, high above, stout limbs hairy with a green brush of needles. Newly arrived here—city kids enchanted with the beauty of the forest—my new bride and I spread a blanket on the ground beneath that huge tree and made tender love among its roots. Ah, youth!

Knowing what I know now, we should have worn hard hats. For what we did not realize in our reckless ecstasy was that the ancient tree was already dying. Some eighty years earlier, a pair of burly loggers had likely stood on the same ground, peered up at the tree and known that its heart was not sound.

"There were different signs that told you," explains Marcus Isbister. Born on this little island in 1924, Marcus remembers when logging crews were still cutting virgin forest here. "One sign you looked for was what they called conk blossoms, which are little funguses growing on the trunk. There was also what they called a blind conk that was inside a lot of the trees—it didn't show any blossom, but sometimes there'd be a bulge on the trunk."

In those days any tree in less than prime condition wasn't worth cutting. Not only were the huge trees a tremendous amount of work to bring down by axe and cross-cut saw, but also if you felled one that you couldn't sell, it was a damned nuisance to fall and skid other logs around. Boomed and towed to Vancouver, prime logs, says Marcus, fetched about three dollars per thousand board feet. "That was for number-one logs!" he laughs. "So anything that was the least bit doubtful they didn't bother with."

I'm glad they didn't. The big fir that died last June was one of these rejects. At first I felt saddened by its death, as though it were somehow gone and I should miss it as one misses a dead friend. That's a nice sentiment, but a bit misguided. For this great tree was already stout before Shakespeare had written a sonnet, and likely will continue to feed new life with its flesh centuries after you and I are forgotten.

Not a true fir, a Douglas-fir (*Pseudotsuga menziesii*) typically lives about 750 years and may survive as long as 1200 years. But even that enormous lifespan is only half the tale. Forest ecologists Jerry Franklin and Miles Hemstrom have concluded that a large Douglas-fir exercises influence on its growing site for as long as 2,000 years! The first millenium is as a growing tree, 500 more years as a dead tree, part of that as a standing snag, and another 500 years as a seedbed for succession species. I need shed no tears for the big tree I see from my window. It has merely entered a different stage in its forest role—one which will continue vibrating with new life long after I'm dead and gone.

* * * * *

For years foresters and forest workers dismissed dead and dying trees as "decadent," "culls" and "snags"—useless and dangerous nuisances to be cleared away in order to make room for a new, vigorous and productive forest. Forest companies, which rely upon a steady supply of old-growth timber for healthy profit margins, still talk that way. But in recent years forest ecologists and holistic foresters have come to recognize that large, dead trees—both standing and fallen—far from being a wasted resource, are one of the most important components of a forest ecosystem.

Discussing forests generally, Charles Elton writes in *The Pattern of Animal Communities*: ". . . dying and dead wood provides one of the two or three greatest resources for animal species in a natural forest. . . . if fallen timber and slightly decayed trees are removed the whole system is greatly impoverished of perhaps more than a fifth of its fauna." Environment officials point out that dozens of species of birds, mammals and amphibians depend upon these "wildlife trees."

But that total pales almost to irrelevance when compared with the untold thousands of micro-organisms whose work it is to break down dead wood and leaf litter and eventually transform their components into soluble compounds upon which the forest may feed. Science is only now beginning to scratch the surface of understanding the processes of decomposition in temperate rain forests, but researchers continue to be astounded at the diversity and complexity they're finding. A recent *New York Times* article on the linchpin role of microscopic arthropods in forest soil quoted experts as identifying 3,400 different arthropod species at an Oregon research forest, and estimating that as many as 8,000 different species might be present. Surveys have shown that a single square metre of forest soil can swarm with as many as 200,000 mites from a single suborder, along with untold thousands of other decomposer species. The soil in a temperate rain forest, the studies suggest, contains what the *Times* calls "some of the most explosive biological diversity found on earth."

* * * * *

The conk blossoms and blind conks spotted by early fallers were actually the fruiting bodies of wood-destroying fungi. These heart rots are a major cause of death in conifers. In this tree they'll now be joined by other fungi and bacteria which will lead a massive invasion of parasites and predators whose work will, over several centuries, transform the tree into a soft mound of humus on the forest floor. A tree's thick outer bark is its only defence against these deadly invaders, and wood-destroying fungi typically gain entrance to living trees through wounds caused either by falling trees or by branches having snapped off, or through bark beetle holes.

"Like looters in a beseiged city," is how Missouri naturalist James Jackson characterizes the assault on a dying tree in his book *A Biography of a Tree*, where he describes the final years in the life of a white oak.

Several types of fungi attacked the oak. After entering bacteria-softened scars, they spread inwardly by pallid, microscopic, ever-branching threads wherever moisture allowed, releasing enzymes along the way which digested cellulose fibres and their cementing lignin.... slowly, endlessly, they found pathways of least resistance through softer, larger-celled spring wood before penetrating the intervening rings of harder summer wood.

A tree's inner bark and cambium hold the highest concentrations of proteins, and it is here that the first opportunistic invaders take up residence. In turn the sapwood, then the heartwood and lastly the bark will yield to different hordes of decomposers. Often a fungus's relentless penetration works in tandem with other organisms: nitrogen-fixing bacteria soften the wood and make nitrogen available to the fungus; boring and chewing beetles carry the spores of fungi into their galleries which serve as moist, warm fungus incubators.

One whole group of bark beetles, called ambrosia beetles, bore tunnels primarily through the sapwood of sound trees. But they don't eat the wood itself; they cast the powdery borings out of their holes, and feed instead on ambrosia fungi which grow along the beetle galleries. Other beetles, like the larvae of the brilliantly green golden buprestid, bore right into the heart of the tree, thus opening it to deep fungal penetration. The female beetle deposits her eggs in cracks or crannies on the tree's bark. When larvae hatch from the eggs, they start chewing their way into the tree. The larvae might tunnel away for two years or more, sometimes excavating a cavity up to five metres long! The larvae eventually work their way to near the surface where they'll pupate and emerge in the spring as adults. Still other armies of beetles prey on the wood borers. The larvae of checkered beetles seek out and eat the eggs and larvae of wood-boring beetles in their tunnels. Tiny blister beetles devour the eggs and larvae of other beetles found under the bark.

Experts say there are many hundreds of insect species directly associated with deteriorating trees, each of them operating in a complex symbiosis of predators, parasites and scavengers. Wood-chewing insects include beetles, ants, termites and wood-tunnelling mites. Certain other insects—collembolans, mites and ambrosia beetles—graze on micro-organisms. Predator mites, spiders, fierce pseudoscorpions and centipedes prey on the smaller insects. As decomposition continues, another whole army of detritus-eaters—earthworms, mites, millipedes, isopods and earwigs—take up residence as well. Eventually the softened snag will fall and become home to sowbugs, snails, slugs and salamanders and the enigmatic slime molds—those yellowish strands of naked protoplasm that ooze over and into soggy wood, sopping up bacteria as they go.

Perhaps no creature better exemplifies the complex patterns of symbiosis within a rotting bole than the dampwood termite. On warm summer evenings, a mated king and queen termite, after a clumsy nuptial flight, alight and are drawn to sound but soggy wood by the presence of special acids and aldehydes produced by certain decomposer fungi. There the royal pair establishes its colony, which will grow into a complex social organization marked with completely cooperative behaviour from all members. While soldier termites with specially developed heads and jaws stand guard over the colony, an obedient under-class of workers digs and burrows new galleries, feeds the royal pair and their nymphs, grooms the young and cleans their nurseries. Writes Frances L. Behnke in *A Natural History of Termites*, "The complicated termite community functions so smoothly as a cooperative unit that the colony resembles a giant superorganism."

The workers, writes Behnke, employ sawlike cutting edges on their jaws to shave off wood particles. Strong jaws grind the shavings, mixing them with saliva. Passed from crop to gizzard to upper and lower intestine, the cellulose fibres are attacked by digestive secretions and enzymes. Tens of thousands of protozoa, living in the intestines, secrete enzymes which engulf each wood particle and convert about half its cellulose into sugars which are then available to the animal as energy source. Chewing its intricate chambers, a termite's body—basically unchanged in the last 50 million years, so successful is it at its work—functions like a tiny anaerobic chamber munching its way through the rotting log!

* * * * *

Unfortunately, the thousands of organisms that work to break down cellulose and thus sustain a forest make no distinction between natural woody debris and wooden buildings or other human constructions. The more visible of the decomposer organisms—beetles, carpenter ants and termites—are seen by anxious home-owners not as tremendously beneficial components of a forest ecosystem, but as destructive nuisances, subversive pests determined to pull one's house down about one's ears.

In our part of the world, moisture greatly accelerates the work of the wood-decomposing community, particularly the fungi. My pesticide handbook notes that "even when wood decay is initiated by insects, the greater damage is usually done by saprophytic fungi, which are either directly introduced by the insects, or find the abandoned insect galleries a suitable environment."

But even in a rain forest climate, wood decay can be greatly retarded by proper building procedures. By excluding moisture, one excludes many of the most destructive organisms, including dampwood termites. Roofs should have a generous overhang and adequate downspouting to avoid splash, as well as appropriate flashing around chimneys, skylights and other joints. Vapour barriers, ventilation of attics and crawl spaces, and proper drainage must be included in building design. Green wood should never be used, and nowhere should wood touch soil. Scrap wood should be removed from around the building, and repairs should be made promptly to leaking roofs or plumbing.

There are many chemical wood preservatives on the market, but these are among the most toxic substances available commercially, and must be used with extreme discretion. Some are formulated to kill insects, others fungi, and some both. Creosote, dieldrin and tetrachlorophenol are three that are commonly sold to combat decay, fungi, sap stains, moulds, beetles and termites. "Wood preservatives are toxic to mammals, birds and fish," warns my pesticide handbook. "Many are very hazardous and may cause eye injury and skin irritations. Protective gear is mandatory."

* * * * *

So too, I think, is a change of attitude. Frances Behnke writes that termites are usually thought of as a destructive force, but that they play a vital role in a beneficial ecological recycling system: "Without them and other decomposers the world would smother in its debris, and all life would come to an end." Instead, the reverse happens, and the decomposers underpin a tremendous diversity of other life forms, both plant and animal. The enormous concentration of insects in a decomposing snag in turn attracts insectivorous birds, particularly woodpeckers. The Douglas-fir in our yard is already pocked in places with small, round holes—probably the work of hairy or downy woodpeckers. They'd be after carpenter ants and beetles, their tunnels revealed by a dust of powdery castings on the outer bark. Big pileated woodpeckers, whose maniacal laughter echoes through the springtime woods, chisel large squarish holes in snags to pluck out ants, termites and beetle larvae. Red-breasted sapsuckers and flickers also chip their way into insect galleries. Some snags are so rich in insect pickings they're called "candy trees" for birds.

Conveniently, these woodpeckers are also cavity-nesters, and in excavating nest sites, they provide subsequent dwelling and nesting opportunities for other species. "Few places are more eagerly sought than an abandoned woodpecker cavity," writes James Jackson. Many bird species, including certain owls and ducks, are cavity nesters that are entirely dependent on tree holes. As well, bats, raccoons and small mammals gladly make use of cavities which are high enough to provide safety from predators and deep enough to be thermally regulated. The cavities also serve as entryways for yet more decomposer organisms to directly attack the tree's heartwood. Then the accumulated excrement of successive cavity dwellers creates new opportunities for decomposition.

Wildlife biologists list a whole range of functions which snags serve for creatures as large as black bears and as small as pygmy nuthatches. The snag out our front window is a favourite perch for a pair of impertinent ravens who will sit on a topmost branch gossiping and griping for hours. In a more refined manner, bald eagles also favour the tree as a perch and lookout post. Turkey vultures and ospreys are listed as other large birds which use old fir snags for loafing sites as well as hunting and hawking perches. Woodpeckers roost on them and also use snags for communication drumming and courtship location. Ospreys, eagles and hawks build their enormous platform nests in the topmost branches, while certain bats, creepers and small mammals nest in cavities under the thick bark. So while countless micro-organisms and insects are chewing their way through the tree's dead heart, a similar array of larger creatures is feeding, nesting, resting and overwintering on the snag.

Researchers now recognize what forest dwellers have known since time immemorial: that fallen trees are a major source of soil organic material and of essential nutrients that are released through the decomposition process. Once sufficiently decayed, a fallen tree becomes a nurse log for other plants, again in a complex pattern of symbiotic relationships.

In the woodlands around our place, plants such as western hemlock, huckleberry, western red cedar and salal are the opportunists that colonize nurse logs. Some of these establish a symbiotic (mycorrhizal) association with other live plants, particularly fungi, rooted in the rotting wood. The fungi inoculate the roots of certain tree species, and eventually grow to form extensions of a tree's root system. The fungus supplies its host with nutrients absorbed from the soil, and the tree supplies its fungal partner with sugars and other products of photosynthesis. Small mammals and decomposer invertebrates living in and around the rotting logs play an important role in dispersing fungal spores. It is this vital connection between decomposing wood, mycorrhizal fungi, mammals and other fungus eaters, and healthy new trees—popularized by writers such as Chris Maser, author of *The Redesigned Forest*—that is now promising to reform human mismanagement of forest lands.

When I walk through the dense conifer woods near our home I feel myself blessed to be in a forest rich in decomposition. Thanks in part to the discerning eye of those loggers working in my grandparents' time, these woods still boast some old Douglas-fir in the declining years of life. And there are dozens of large snags in various stages of collapse—pitted with woodpecker holes, great sheets of bark hanging loosely from their sides while mounds of soft, brown debris accumulate at their enormous bases. One of my favourites is an old Douglas-fir broken off about ten metres up in the air and topped with an unruly mop of sword ferns and hemlock seedlings.

Some old logs are freshly fallen onto the forest floor, while others have lain there so long they're now nothing more than soft, moss-covered pillows of powder on the ground, held together by the long, stringy roots of new trees feeding from the old. And there, looming above them all, is the just-dead Douglas-fir, its dominance ended, beginning its lengthy epoch of collapse.

Walking among these decomposing beauties, touching their coarse surfaces, seeing the shelflike sporophores of fungi blossoming along their bodies, hearing the hysterical jackhammering of woodpeckers and smelling the rich, earthy aroma of decay—how can I not rejoice in a reassuring sense of place. How not feel at least a little humble amid the vast and wonderful complexity of a timeless forest.

Metric Conversion Table

Length

1 inch	2.54 centimetres
1 foot	30.48 centimetres
1 yard	0.914 metre
1 mile	1.609 kilometres

Area

1 square yard	0.836 square metre
1 acre	0.405 hectare
1 square mile	2.590 square kilometres

Volume

1 cubic inch	16.837 cubic centimetres
1 cubic foot	0.028 cubic metre
1 cubic yard	0.765 cubic metre

Weight

1 ounce	28.35 grams
1 pound	0.454 kilogram
1 short ton	0.907 metric tonne
1 long ton	1.016 metric tonnes

Bibliography

General

British Columbia, Ministry of Environment. *Handbook for Pesticide Applicators and Pesticide Dispensers.* Victoria: 1979.

Evans, Ivor H. *Brewer's Dictionary of Phrase and Fable.* London: Cossell Ltd., 1981.

Forsyth, Adrian. *A Natural History of Sex.* New York: Charles Scribner and Sons, 1986.

Frazer, J.G. *The Golden Bough.* New York: MacMillan, 1942.

Funk and Wagnalls Standard Dictionary of Folklore, Mythology and Legend. New York: Funk and Wagnalls, 1972.

The Guiness Book of Animal Facts and Feats. Middlesex: Guiness Superlatives Ltd., 1982.

Harrowven, Jean. *Origins of Rhymes, Songs and Sayings.* London: Kaye and Ward, 1977.

Larousse Encyclopedia of Animal Life. New York: Hamlyn Publishing Group, 1967.

Nilsson, Greta. *The Endangered Species Handbook.* Washington, D.C.: Animal Welfare Institute, 1983.

Rood, Ronald. *It's Going to Sting Me!* New York: Simon and Schuster, 1976.

Street, Philip. *Animal Reproduction.* New York: Taplinger Publishing, 1974.

Vogel, Virgil J. *American Indian Medicine.* New York: Ballantine Books, 1973.

Yepsen, Roger B., editor. *Organic Plant Protection.* Emmaus, Pa.: Rodale Press, 1976.

Amphibians and Reptiles

Autotte, Christian. "Seeing Is Believing." *Nature Canada.* Summer, 1988.

Beebee, Trevor. *Frogs and Toads.* London: Whittet Books, 1985.

Brodie, Edmund D. "Genetics of the Garter's Getaway." *Natural History.* July, 1990.

Carl, G. Clifford. *The Amphibians of British Columbia.* Victoria: Royal B.C. Museum, 1966.

———. *The Reptiles of British Columbia*. Victoria: Royal B.C. Museum, 1968.

Gregory, Patrick T. and Campbell, Wayne R. *The Reptiles of British Columbia*. Victoria: Royal B.C. Museum, 1984.

Hall, Nor. *The Moon and the Virgin*. New York: Harper and Row, 1980.

Lillywhite, Harvey B. "Snakes Under Pressure." *Natural History*. November, 1987.

Phillips, Kathryn. "Frogs In Trouble." *International Wildlife*. November/December, 1990.

Birds

Angell, Tony. *Ravens, Crows, Magpies and Jays*. Vancouver: Douglas and McIntyre, 1978.

Beebe, Frank L. "A Study in Character—The Story of a Raven." *Victoria Naturalist*. Victoria, 1957.

Brenchley, Anne. *Ravens and Their Effects on Sheep Predation on Saltspring Island, British Columbia*. Victoria: Ministry of Agriculture and Food, 1985.

Bruemmer, Fred. "Ravens: Smartest Birds in the World." *International Wildlife*. September/October, 1984.

Dillard, Annie. *Pilgrim at Tinker Creek*. New York: Harpers Magazine Press, 1974.

Field Guide to the Birds of North America. Washington, D.C.: National Geographic Society, 1983.

Goodwin, Derek. *Crows of the World*. Seattle: University of Washington Press, 1987.

Heinrich, Bernd. "The Raven's Feast." *Natural History*. February, 1989.

Kennedy, Des. "The Great Transformer." *Nature Canada*. Summer, 1988.

Morris, Rev. F.O. *British Birds*. London: Peerage Books, 1981.

Stefferud, Alfred, editor. *Birds in Our Lives*. Washington, D.C.: Department of the Interior, 1966.

Stokes, Donald. *A Guide to Bird Behaviour*. Boston: Little, Brown and Co., 1979.

Willet, George. "Variations in North American Ravens." *Auk*. 1941.

Insects and Molluscs

Behnke, Frances L. *A Natural History of Termites*. New York: Charles Scribner and Sons, 1977.

Crompton, John. *The Life of the Spider*. New York: New American Library, 1954.

Dondale, Dr. C.D. "The Tie That Binds." *Nature Canada*. Summer, 1988.

Fabré, J. Henri. *The Life of the Spider*. New York: Dodd, Mead and Co., 1914.

Forsyth, Adrian. "Stung." *Harrowsmith*. July/August, 1991.

Hansell, Michael H. "Wasp Paper-Mache." *Natural History*. August, 1989.

Kennedy, Des. "Nature's Slimy Recyclers." *Nature Canada*. Summer, 1990.

Lang, Aubrey. "Masters of Survival." *Canadian Geographic*. February/March, 1990.

Lehane, Brendan. *The Compleat Flea*. London: John Murray Ltd., 1969.

Oldroyd, Harold. *The Natural History of Flies*. London: Wiedenfeld and Nicholson, 1964.

Pitcairn, Richard and Susan H. *Natural Health for Dogs and Cats*. Emmaus, Pa.: Rodale Press, 1982.

Rollo, C. David. "The Behavioral Ecology of Terrestrial Slugs." PhD thesis (unpublished). University of British Columbia, 1978.

Runham, N.W. and Hunter, P.J. *Terrestrial Slugs*. London: 1970.

Swan, Lester A. *Beneficial Insects*. New York: Harper and Row, 1964.

Westcott, Cynthia. *The Gardener's Bug Book*. New York: Doubleday, 1946.

The Yellowjackets of America North of Mexico. Washington, D.C.: Department of Agriculture, 1981.

Mammals

Banfield, A.W.F. *The Mammals of Canada*. Toronto: University of Toronto Press, 1974.

Boland, Maureen and Bridget. *Old Wives' Lore for Gardeners*. New York: Farrar, Straus and Giroux, 1976.

Burton, Maurice and Robert. *Encyclopedia of Mammals*. London: Octopus Books, 1975.

Caras, Roger A. *North American Mammals*. New York: Galahad Books, 1967.

Calhoun, J.B. *The Ecology and Sociology of the Norway Rat*. Baltimore: U.S. Public Health Service, 1963.

Carl, G.C. and Guiget, C.J. *Alien Animals in British Columbia*. Victoria: B.C. Provincial Museum, 1972.

Harris, Joan Ward. *Creature Comforts*. Toronto: Collins, 1979.

Hill, J.E. and Smith, J.D. *Bats: A Natural History*. Austin: University of Texas Press, 1984.

Lorenz, Conrad. *On Aggression*. New York: Harcourt, Brace and World, 1966.

North, Sterling. *Raccoons Are the Brightest People*. New York: Dutton and Co., 1966.

Nowack, Ronald M. and Paradiso, John L. *Walker's Mammals of the World*. Baltimore: Johns Hopkins University Press, 1983.

Plants

Berglund, B. and Bolsby, C. *The Edible Wild*. Toronto: Ptarmigan Press, 1971.

Bland, John. *Forests of Lilliput*. Englewood Cliffs, N.J.: Prentice Hall, 1971.

Canada, Department of Northern Affairs and National Resources, Forestry Branch. *Native Trees of Canada*. Ottawa: Queen's Printer, 1956.

Collingwood, G.H. and Brush, Warren D. *Knowing Your Trees*. Washington, D.C.: American Forestry Association, 1960.

Coon, N. *Using Wayside Plants*. Waterton, Mass.: Eaton Press, 1957.

Crockett, Lawrence, J. *Wildly Successful Plants*. New York: MacMillan Publishing, 1977.

Daar, Sheila. *Integrated Weed Management for School Grounds*. Napa, Calif.: JMI Inc., 1982.

Frazer, J.D. and Evans, S.A. *Weed Control Handbook*. Dorking: Adlard and Son, 1968.

Gordon, Lesley. *A Country Herbal*. New York: W.H. Smith, 1980.

Hatfield, Audrey Wynne. *How to Enjoy Your Weeds*. New York: Collier Books, 1973.

Houghton, Claire Shaver. *Green Immigrants*. New York: Harcourt, Brace, Janovich, 1978.

Hunter, Beatrice Trum. *Gardening Without Poisons*. New York: Berkley Medallion Books, 1964.

Jackson, James P. *The Biography of a Tree*. New York: Jonathon David Publishers, 1979.

Jason, Dan and Nancy et al. *Some Useful Wild Plants*. Vancouver: Talonbooks, 1972.

Kennedy, Des. "Death of a Giant." *Nature Canada*. Spring, 1991.

Lust, John B. *The Herb Book*. New York: Bantam Books, 1974.

Maser, Chris. *Forest Primeval*. Toronto: Stoddart Publishing, 1991.

―――. *The Redesigned Forest*. San Predro, Calif.: R. and E. Miles, 1988.

Schofield, Wilf B. *Some Mosses of British Columbia*. Victoria: Royal B.C. Museum, 1973.

Spencer, Edwin Rollin. *All About Weeds*. New York: Dover, 1974.

Szczawinski, Adam F. and Hardy, George A. *Guide to Common Edible Plants of British Columbia*. Victoria: Royal B.C. Museum, 1962.

Tarrant, Robert. "Some Effects of Alder on the Forest Environment." *Biology of Alder: Proceedings of a Symposium of the Northwest Scientific Association*. Corvallis, Ore.: U.S. Department of Agriculture, Forest Service, undated.

Turner, Nancy J. *Food Plants of British Columbia Indians*. Victoria: Royal B.C. Museum, 1975.

Walstad, John D., editor. *Forest Vegetation Management for Conifer Production*. New York: John Wiley and Sons, 1987.

Wyman, Donald. *Wyman's Gardening Encyclopedia*. New York: MacMillan, 1971.

Index

A

Abortion, spontaneous, 168
Alder: description of, 17; die-back of, 19; as firewood, 21; in forest ecology, 19; growth characteristics of, 17, 18, 21; industrial uses of, 21-22; medicinal properties of, 16; native uses of, 16, 21; range of, 15; red, 14, 15, 16, 17, 18, 20; seeds of, 18; site domination by, 19; used in site stabilization, 19-20; soil improvement and, 19; species of, 14-15; traditional uses of, 20, 30; as wildlife habitat, 20
American Aboriginals: Apache, 59; Algonquin, 116; Aztec, 159; Cherokee, 148; Chippewa, 159; Haida, 85, 183; Hopi, 147; Huichal, 152; Kwagulth, 34, 148; Mayan, 8; Nuu-chah-nulth, 218, 148; Seminole, 148; Yakut, 147
American harvest mice, 94
Ants, carpenter, 199, 200, 201
Appophalation, 50
Arachnida, 127
Arakunem, 116, 117, 122
Arthropods, 197
Audubon Society, 65, 70, 137
Australia, 98-99, 172, 182
Australian aboriginals, 11, 147
Avicides, 70, 140
Avitol, 70

B

Bacillus thuringiensis, 80
Bacon, Francis, 109
Bacteria, 198
Ballooning, by spiderlings, 134
Banana slug. *See* Slugs
Basking, by snakes, 149, 151
Bats: big brown, 4, 5, 7; biology of, 3; breeding and reproduction by, 6; care of young by, 6-7; diseases of, 9; distribution of, 3-4; use of echolocation, 4-5; extermination of, 2, 10; fear of, 2-3, 7; flight of, 3, 4; guano, 6, 9; hibernation by, 5-6, 7; homing instinct in, 5; hunting methods of, 4-5; legislation concerning, 10; life span of, 7; little brown, 4, 6, 7; myths about, 7-8; origin and evolution of, 3; rabies in, 8, 9, 187; repellents of, 10; senses of, 3; species of, 3-4; vampire, 8
Beebe, Frank, 109
Beebe, Trevor, 177, 179, 180, 181
Beetles, 198-99, 200, 201
Behnke, Frances L., 199, 201
Bertolotto, Signor, 24, 25
Black Death, 26, 27
Black slug, 48
Black widow spiders. *See* Spiders
Blake, William, 26
Bland, John, 84, 86
Blaustein, Andrew, 181
Bluebottles, 76, 77
Boland, Maureen and Bridget, 39, 40, 41
British Crop Protection Council, 61
Brodie, Edmund D., 151-52
Browne, William, 15
Brown, Tom, 176
Brown rat. *See* Rats
Bubonic plague, 27
Bufotinine, 181
Bullfrogs, American, 174, 182

C

Calhoun, J.B., 167-68
Campbell, Thomas, 161
Campbell, Wayne, 175, 177, 178
Carter, Alan, 51
Cat fleas, 28
Cavity nesters, 201
Childs, Dr. James, 170

China, 11, 27, 56, 163-64
Celsus, Cornelius, 30
Cicero, 111
Clearcut logging, 18, 22, 38, 43-44, 182
Commensal species, 95, 165, 166, 172
Congo floor maggot, 81
Conifer suppression, 19
Conks, 196, 198
Convention on International Trade in
    Endangered Species, 152
Corvidae, 106-07
Craighead, Frank, 107
Cryptococcosis, 69
Culpeper, Nicholas, 57, 158, 159

D
Dandelion: blanching of, 57; as
    companion plants, 63; as coffee
    substitute, 57, 59; culinary uses of,
    57; history of, 56-57; control of,
    61-63; medicinal uses of, 57-59;
    names for, 56-57, 58; propagation
    by, 60-61; Society, 60; vitamin
    content of, 58; as weeds, 61; wine,
    57, 58; use of wild bees, 63
DDT, 10, 11, 31, 79
Deer: breeding, 42; browse, 38;
    Columbian blacktail, 36, 37, 38, 41,
    43; deer-proof plants, 39; description
    of, 34, 35, 36; dogs and, 42-43;
    fawns, 35, 36-37, 41, 43; fences
    against, 40-41; hunting of, 42,
    43-44; mortality of, 42, 43; moult
    of, 34, 41; mule, 36, 38; parasites
    of, 43; as pests, 38, 39; predators
    of, 43; repellents of, 39-40; rut of,
    41-42; size of, 36; species of, 36;
    white-tailed, 36, 38; wolves and,
    43-44
Diatomaceous earth, 32, 53
Diazanon, 193
Dickinson, Emily, 146
Dillard, Annie, 140
Distemper, canine, 120
Disney, Walt, 35
Dog fleas, 28
Donne, John, 26
Douglas-fir, 19, 20, 38, 43, 195, 196,
    197, 201, 202, 203

E
Earle, Alice Morse, 61
Echolocation, 4-5
Eibl-Eibesfeldt, J., 98
Elton, Charles, 197
Encephalitis, 69
Endangered Species Act, 10
Evelyn, John, 26

F
Fabré, J. Henri, 129, 130, 131, 132,
    133, 134
Falcons, peregrine, 70
Fawns. See Deer
Fernald, M.L., 57
Fleas: bites of, 26, 28; biology of, 28;
    cause of Black Death, 26-27; circus,
    25; collars for, 31; common, 25, 28;
    control of, 31-32; copulation, 28-29;
    development of, 29; as disease
    vectors, 94; in folklore, 25, 26,
    29-31, 74; jumping ability, 25-26;
    lifespan, 29; performing, 24-25; in
    raccoon dens, 118; repellents, 29-31;
    on rats, 27; species of, 28
Flies: benefits of, 81; blow, 77, 79, 80,
    81; control of, 79-80; as disease
    vectors, 78-79; feeding by, 76; flesh,
    76-77, 79; house, 76, 77, 78, 79,
    80; maggots of, 74-75, 76, 77, 80,
    81; on mosses, 88; pigeons and, 69;
    plagues of, 78; reproduction by, 77,
    78; species of, 75-76, 80-81
Forsyth, Adrian, 50, 60, 132, 133, 168,
    186-87, 192, 193
Franklin, Jerry, 197
Frazer, Sir J.G., 11, 100, 101, 147, 148,
    152, 172
Frazier, C.A., 187
Frogs, species of, 174-75. See also
    Toads and frogs
Fungi, 198, 199, 200, 202, 203
Fullard, Dr. James, 5

G
Gametophyte, 86, 87
Galloway, Paul, 69
Gastropoda, 47
Gay, John, 111

Gemmae, 86
Gilgamesh, Epic of, 109
Gordon, Lesley, 57, 58, 61, 157, 158, 159, 160, 161
Goodwin, Derek, 109
Great Corbie-crow, 106
Great slug of Europe, 48, 50
Great Transformer, 106, 113
Greenbottles, 77, 179
Green, David, 175, 177, 178
Grey garden slug, 48
Guano bats, 6
Guano, of bats, 9
Gypsies, 11

H
Haida Gwaii, 38, 85, 89, 106
Hall, Nor, 146
Hansell, Michael, 188, 189, 190, 191
Harris, Joan Ward, 37, 117, 119, 121
Harrowven, Jean, 12
Hatfield, Audrey Wynne, 161
Haughton, Claire Shaver, 63
Health and Welfare Canada, 62
Heinrich, Bernd, 107, 108, 111
Hemstrom, Miles, 197
Hennepin, Fr. Louis, 176
Herbicides, 18, 62, 90, 156
Hibernation: of bats, 5-6, 7; of raccoons, 118
High plains grasshopper, 114
Hill, John, 3, 7, 9, 11
Histoplasmosis, 9
Hobson, R.P., 77
Hollander, W.F., 69
Homing instinct, 5, 67-68, 177
Hounds, coon, 122
Hunter, Beatrice Trum, 40, 80, 120, 160, 161, 171
Hunter, P.J., 47, 51, 52, 53
Hunting spiders. *See* Spiders

I
India, 27, 152
Indians, American. *See* American Aboriginals
Indian flying fox, 7
Indian spinach, 157
Ireland: rats in, 171; snakes in, 146;
    toads in, 176
Irish moss, 86, 90

J
Jackson, James, 198, 201
Japan, 19, 60, 90
Jaundice, 166
Jonson, Ben, 171
Josselyn, John, 16

K
Kern, Stephen, 10
Kinsey, A.C., 57

L
Larvadex, 79-80
Lehane, Brendan, 24, 25, 27, 28, 29, 30
Lillywhite, Harvey, 150
Linnaeus, 116
Liverworts, 85, 86, 89
Longfellow, Henry Wadsworth, 86
Lorenz, Konrad, 98, 109, 169, 170, 172
Lupton, Thomas, 176

M
Maggots, 29, 179. *See also* Flies
Marlowe, Christopher, 24, 111
Maser, Chris, 202
McDonald, Elvin, 90
Metamorphosis, 77, 179
Mexican free-tailed bats, 11
Mexican marigold, 80
Mice: aggression in, 107-08; benefits of, 102-03; control of, 100-02; damage caused by, 100-01; deer, 94, 97, 98, 100, 102, 103; eating and sleeping patterns of, 98; fear of, 98; folklore about, 95, 96; house, 94, 95, 97, 98, 103; medicinal uses of, 102-03; migrations of, 95; nests of, 97-98, 100; plagues of, 93-94, 98-99; predators of, 99, 102; reproduction by, 97; sayings about, 96-97; space, in, 103; species of, 94-95; Super, 102; territory of, 98; white, 96; witches and, 95, 103
Milky slug, 48
Mollusca, 47, 48
Moodie, Susannah, 59

Morris, Rev. F.O., 111, 113, 136, 141, 142, 143
Moss: biology of, 86; description of, 84, 85; disjunct, 89; endemic, 89; evolution of, 86; feather, 84; Irish, 86, 90; killer for, 83, 84, 90; lawns of, 90; lawns, in, 89-90; luminous, 88; names of, 86; Pacific Northwest, in, 84; reproduction by, 86, 87; Scottish, 86, 90; sphagnum, 88; spores of, 87; substrates used by, 88
Moths, 5
Moufet, Thomas, 30
Mule deer. *See* Deer
Muridae, 94
Murinae, 94
Murine typhus fever, 166
Myophobia, 98

N
Native Americans. *See* American Aboriginals
Native herbalists, 59
Newcastle disease, 69
Nimpkish Valley, 43-44
Nitrogen fixation, 19
Norse mythology, 108
North, Sterling, 115-16, 117, 121, 122, 123
Northwest garter snake, 149, 151, 152
Norway rat. *See* Rats
Nurseryweb spiders, 128

O
Oldroyd, Harold, 75-80
Olfactory coercion, 168
Orb weavers, 126, 129, 130, 133
Oregon alder, 14
Oriental rat flea, 25, 27
Ovid, 24, 109

P
Pack rat, 165
Paradichlorobenzene (PDB), 10
Parthenogenesis, 60-61
Pasteurella pestis, 27
Patrick, Saint, 146, 147, 148, 153, 157, 176
Penant, Thomas, 176
Pennsylvania Germans, 58, 103

Pepys, Samuel, 158
Pesticides, 11, 40, 60, 79, 181
Phenoxy acid compounds, 62
Phoneutria spiders, 131
Pigeons: brain of, 68; control of, 69-70; in cities, 69-70; diseases of, 69; displays by, 71; droppings of, 69; in Egypt, 66; game, 67; homing in, 67-68; lobby, 70; milk, 72; names for, 66; nesting by, 71-72; pair, 71; parasites of, 69; passenger, 72; as pests, 69; pie, 72; racing, 67; sayings about, 68; shooting of, 71; show, 67; squabs, 66-67; stupidity of, 68; survival rates of, 72; vision in, 68; use in war, 68; wood, 72
Pitcairn, Richard and Susan, 31
Pit-lamping, 42
Pliny, 30, 66, 67, 103
Poe, Edgar Allan, 111
Poria Weirii, 20
Psittacosis-ornithosis, 69
Puparium, 77-78
Pythons, 146, 151, 152

Q
Queen Charlotte Islands, 38, 85, 89, 90, 106, 123
Queletox, 70

R
Rabies, 8, 9
Raccoons: adaptability of, 116, 119; breeding and reproduction by, 118; caps and coats made from, 122, 123; communication by, 117; as control predators, 123; dens of, 118, 119; farming of, 123; food of, 118-19; hibernation of, 118; hunting of, 122; names for, 116; native uses of, 122; as pets, 121; as pests, 119-20; predators of, 118; range of, 117; repellents of, 120; senses of, 117; size of, 117; species of, 116; trapping of, 120; as trickster, 116
Rats: aggression in, 168, 169, 170; attacks on humans by, 167; birth control by, 168; burrows of, 167; in China, 163-64; colonies of, 167-68,

169, 172; commensal, 165, 166, 172; control of, 164, 171; damage to foodstuffs by, 166; densities of, 167, 168; as disease vectors, 166; exorcisms of, 171; extinctions of, 172; fleas and, 27; food of, 166-67; information transmission by, 170-71; Irish, 171; laboratory, 172; meat of, in China, 163-64; names for, 165; Norway, 165, 166; Old World, 165; pigeons and, 69; poisons for, 171; reproduction by, 167, 168; roof, 165; sayings about, 164; Super, 171; survival of, 168, 170; white, 172
Ravens: adaptability of, 107, 110; in Arctic, 110; as carrion eaters, 111-12; communication by, 107, 108, 109; in Christian mythology, 109; as omen of death, 111; as predators of domestic livestock, 112-13; dominance in, 108; folklore about, 108-09, 110-11; hunting of, 113; intelligence of, 107, 108, 109; lifespan of, 110; mating rituals of, 110; nesting by, 110; in Norse mythology, 108, 109; used for shore sighting, 109; swamping by, 108; as trickster, 106, 113, 116
Red deer, 36
Refugia, 89
Rock dove, 66
Rodenticides, 102, 171
Rodents, 94, 96, 99-100, 168
Roe deer, 36
Rollo, David, 50, 51, 52
Roof rat, 165
Rothschild, Miriam, 29
Runham, N.W., 47, 51, 52, 53

S
Saint Patrick. See Patrick
Salmonella, 69, 120, 166
Sanders, Peter, 18, 20
Sandwell, B.K., 174
Saw fly, 102
Schieffelin, Eugene, 136, 137
Schofield, Dr. Wilf, 84-89
Sex totems, 11
Shakespeare, William, 26, 35, 136, 170, 196

Shamans, 147, 183
Site preparation, 18
Slugs: aggression in, 50, 52; baits for, 53; banana, 46, 47, 48, 51, 54, 149, 161; biology of, 48-49; classification of, 47; control of, 53; evolution of, 47; exotic, 48; feeding patterns of, 51-52; growth stages of, 51; habitat of, 47; mating of, 50; predators of, 52-53; reproduction by, 51; slime produced by, 49-50, 53; snakes and, 52, 149, 153; species of, 47-48
Smith, James, 3, 7, 9, 11
Smith, John, 116
Smith, Joseph, 58-59
Snags, 197, 199, 201, 202
Snakes: basking by, 149, 151; colour patterns and escape behaviour of, 151-52; endangered species of, 152; fear of, 146-47, 153; garter, 148, 149, 151, 152; mythological, 148; native respect for, 147, 148, 152; predators of, 151; protection of, 153; rituals performed with, 147-48; Saint Patrick and, 146, 147; uses of skins, 152; structural adaptations by, 150; swallowing apparatus of, 149-50; venomous, 146
Spalangia endius, 80
Spanish moss, 86
Speckled alder, 14, 15, 20
Spencer, E.R., 63
Sphagnum moss, 88
Spiders: benefits of, 134; black widow, 127, 131, 132, 134, 187; cannibalism by, 132, 133; cocoons of, 133; distinguishing features of, 127; eggs of, 133; eyesight in, 127; fear of, 126; hunting, 127-28, 133; jumping, 127; life cycle of, 134; mating by, 132-33; *Phoneutria*, 131; reproduction by, 133; silk of, 128, 130, 131, 132, 133; species of, 126-27; tarantula, 126, 127, 128, 131-32, 134; venom of, 131, 176; webs of, 126, 129, 130; webspinning, 127, 128-29, 133; wolf, 128, 133; young, dispersal of, 133-34
Spores, 87

Sporophyte, 87
Spotted garden slug, 48
Squabs, 66-67Starlings: benefits of,
   142-43; common, 137, 138;
   communication by, 143; control of,
   140-41; deterrence of, 141-42;
   displacement of native birds by, 138;
   droppings of, 136, 139; eating of,
   140-41; introduction to North
   America of, 136, 137, 139; mating
   by, 138, 142; as pests, 139-40;
   roosts of, 139, 140, 141; shooting of,
   136, 140; slugs and, 53; songs of,
   143; trapping of, 140; in Vancouver,
   137, 142
Stinging Nettles: as companion plants,
   161; in compost, 161; culinary uses
   of, 157-59, 162; description of, 156,
   162; used as fibre, 160-61; medicinal
   uses of, 159-60; miscellaneous uses
   of, 160; superstitions about, 160;
   stings from, 156-57, 162; as weeds,
   157; wildlife dependent upon, 161
Stokes, Donald, 71, 138, 142, 143
Street, Philip, 177, 188, 194
Strychnine, 102, 181
Swamping, 108
Swan, Lester, 80
Swift, Jonathan, 32
Synanthropic flies, 76

T
Tarantula. See Spiders
Taraxacum, 56, 59
Tarrant, Robert, 19, 20
Termites, 199-200, 201
Theraphosid spiders, 126
Thiram, 40
Toads: common, 177, 178-79; licking
   of, 181; marine, 177, 182;
   Northwestern, 174-75, 179; Sonoran
   desert, 181. See also Toads and frogs
Toads and frogs: breeding by, 178;
   consumption of, by humans, 182;
   distinguishing characteristics of, 175;
   evolution of, 176-77; exotic species
   of, 182; eyesight of, 180; habitat
   destruction of, 182; hatred of, 176;
   homing instinct in, 177; migration

of, 177-78; mortality of, 179-80;
   native myths about, 183; plagues of,
   175, 182; population collapses of,
   181-83; reproduction by, 179; skins
   of, 180; songs of, 174-75, 178-79,
   180, 183; species of, 174-75, 177;
   superstitions about, 175-76; toxins
   produced by, 176, 180-81
Tolmie, William, 160
Toxoplasmosis, 69
Trade rat, 165
Tree frogs, Pacific, 174, 175, 177, 178, 179,
   180, 183. See also Toads and frogs
Trickster. See Raven and Raccoon
Tuttle, Dr. Merlin, 10

U
Uroboros, 146
U.S. Coast Guard, 68
U.S. Department of Agriculture, 137,
   187, 192, 194
U.S. National Pest Control Assoc., 97
U.S. Public Health Service, 167

V
Vampire bats, 8
Vespidae wasps, 188
Vikings, 60
Vomit drop, 76

W
Wake, David, 182
Warfarin, 102, 171
Warnock, John, 62
Wasps: attack by, 186; benefits of, 187,
   194; colonies of, 188-89, 192, 193;
   digger, 188; food of, 191; in fruit,
   191; larvae of, 188, 189, 190, 191,
   192; nests of, 188-90, 191, 192, 193,
   194; parasitic, 187; as pests, 191,
   194; population fluctuations of, 191;
   potter, 188; predators of, 188, 190;
   prevention of, 192, 193; queens,
   189, 191, 192, 194; social, 188, 189;
   species of, 187-88; stinger in, 186,
   187, 188, 190, 192; venom of, 176,
   186-87, 188, 193; witches and, 194;
   worker, 189, 191, 192, 193
Wasserman, Dr. Edward A., 68
Weber, Wayne, 112, 113

Web-spinning spiders. *See* Spiders
Weeds, 59, 62-63, 157, 162
Weed trees, 13, 14, 15, 18, 20, 22
Wesley, John, 159
Western harvest mice, 94
White-tailed deer. *See* Deer
Wildlife trees, 197
Witches, 2, 7, 8, 26, 58, 176, 194
Wolf spiders. *See* Spiders
Wolves, 43-44
Wood, decomposers of: attack on tree
  by, 198; dispersal of fungal spores
  by, 202; numbers of, 197-98, 199; as
pests, 200; predators of, 201;
prevention of, 200; symbiotic
relationships of, 198, 199, 201, 202;
value of, 201
Woodpeckers, 201, 202, 203
Wood preservatives, 200-01
Wood rat, 165
Wood, William, 122
Woods, Alex 107-08
World Health Organization, 164, 166

Y
Yellowjackets. *See* Wasps